The Concept of Capital

Bruce R. Scott

The Concept of Capitalism

Professor Bruce R. Scott
Harvard Business School
Boston, MA 02163
USA

ISBN 978-3-642-03109-0 e-ISBN 978-3-642-03110-6
DOI 10.1007/978-3-642-03110-6
Springer Dordrecht Heidelberg London New York

Library of Congress Control Number: 2009931691

Cover design: WMXDesign GmbH, Heidelberg, Germany

Printed on acid-free paper

Springer is part of Springer Science+Business Media (www.springer.com)

Table of Contents

Preface

This monograph on the concept of capitalism is the intellectual core of a larger work, entitled *Capitalism, Its Origins and Evolution as a System of Governance*, due for publication November 2009. The purpose of this monograph is to put forth an original concept of capitalism as a system of governance, including a theory of how it functions at any point in time and how it evolves through time. In the larger book, I present a theory of its origins and evolution and support this theory with a set of country case studies that span both time and geography. It was, in fact, my experience in studying these case studies that led me to the concept presented here as well as to the theory of capitalism's origins and evolution.

In the larger book, I build on the present work, identifying and explaining capitalism as a system of governance for political entities such as nation states. I then supplement these ideas with a description and explanation of three generic economic strategies. Taken together, my studies of economic strategies and specific capitalist systems of governance are intended to enhance and enrich existing literature on "varieties of capitalism". The larger book also includes two appendices; the first explains each of the three levels of capitalism as a system of governance in some detail, while the second focuses on the role of firms as key actors in a modern capitalist system, as illustrated with a US example. Together, the case studies and supplementary appendices will provide readers with a deeper understanding of capitalism as a system of governance, grounded in events both historical and contemporary, albeit with a distinctive emphasis on the experiences of the US from 1630 to 2008.

In publishing this very condensed version of the core theory of the longer book, I would like to acknowledge the support of a few individuals who played unusually direct roles in my work on these two texts. First, I would like to acknowledge the special role of Niels Peter Thomas, my editor at Springer Verlag, in shaping the entire text, including suggestions for the addition of two chapters on the most recent transformations of US capitalism. I would also like to acknowledge the role of Sarah Potvin, my Research Associate from June 2005 to June 2007, for encouraging me to read the writings of Robert Dahl and others on democracy as well as for writing a draft of a chapter on US capitalism in the period 1630-1830. In addition, I would like to acknowledge the role of Linnea N. Meyer, my Research Associate from June 2007 to the present, for writing a draft of a chapter on US capitalism in the period 1830-1937 and for playing a very important role in editing much of the larger book as well as this monograph. I have been very fortunate to have these colleagues and many others as sources of encouragement over these last four years on a project that was initiated in the early 1990s. I have also been fortunate to have had the financial support of the Division of Research at Harvard Business School for both my time and that of a Research Associate over the past 15 years as I composed both this work and the longer book. Finally, I would like to acknowledge the role of Grenelle, my wife, who has been a sounding board, editor, proof reader and source of encouragement on this project for almost twenty adventurous years.

Bruce R. Scott
May 2009

Chapter 1

Introduction

Two systems of governance – capitalism and democracy – prevail in the world today. Both systems are built upon the notion of indirect governance through regulated competition as the key coordinating mechanism among various actors. They can operate simultaneously within the same society because they operate in partially distinct domains; at the same time they can and do influence each other. Indeed, participants in one system can use their positions in that system as a base from which to compete for power in the other.

Competition among economic and political actors is not unique to capitalism and democracy; indeed, the actors of most if not all pairings of economic and political systems of governance are likely to compete for power, as economic resources can be used to purchase political resources, and political resources can be used to generate or distribute economic resources. What makes the power struggle between capitalism and democracy distinct and, I believe, to warrant further examination, is that such use of one system's resources to influence those of the other becomes misuse when held up against the systems' respective principles of free markets and universal political liberties or, more generally, equal opportunity to participate in either system. The goal of leaders in any society characterized by both capitalism and democracy should therefore be to mitigate such misuse of economic or political power, establishing institutions and rules that condition the behavior of actors in each system, economic and political.

However, such a goal cannot be achieved if these leaders and, as it were, society at large, do not understand how these systems function, how they influence each other, or even that they are systems at all. Neither capitalism nor democracy is easily understood and, indeed, there is no standard definition of either. My primary purpose in this work is to provide an original definition and thus a better understanding of capitalism alone, namely: *Capitalism is an indirect, three level system of governance for economic relationships,* as I will explain shortly. I make no claim to provide an original definition of democracy; throughout this work, I assume democracy implies that power in the political system rests with political leaders who are held accountable to a free society by appropriate institutions through periodic elections. I contrast democracy with oligarchy, a condition where significant political power is vested in the economic system and is not necessarily held accountable to anyone.

To understand capitalism as a system of governance is to transcend the boundaries of standard neo-classical economic analysis, moving beyond merely the markets of pure economics to include the institutions and authorities of political economy. Such an understanding requires a more holistic analytic approach, one including insights from political science, sociology and the law. While there could be a number of causes for the apparent difficulties in arriving at a consensus definition of capitalism, it is at least partly due to the question of other attempts being limited by the bounds of a single academic discipline, typically economics.

There is little reason to suppose that defining how a system of governance works should be easy, and a number of distinguished scholars have made important contributions toward such a definition. Adam Smith provided a remarkable insight into how the markets of capitalism can coordinate the actions of literally thousands or millions of people, without any conscious guidance on the part of the quasi-independent economic actors, as they equilibrate supply and demand through the price mechanism. And about a century later, neo-classical economics emerged, with a small group of British economists recognizing that it was the markets that established the values of various good and services, rather than the intrinsic properties of these items.

These discoveries, path-breaking as they were, remain insufficient as a working definition or understanding of capitalism, since they tend to focus on the achievement of market-based equilibrium as though equilibrium, by itself, stands for economic governance. However, the realities of market economies have shown time and again that equilibrium can be achieved in distorted markets where supplies include goods produced by slaves or other forms of forced labor, in speculative bubbles where excessive leverage permits buyers to generate unsustainable levels of demand or supply or, on the contrary, in depressed markets where effective demand is far below a nation's capacity to produce. For equilibrium to be a true reflection of effective societal governance, market prices must reflect true social costs (i.e., factor in the value of the goods and services to society as a whole) and demand must reflect sustainable demand without the use of undue financial leverage by the borrower or the lender (i.e., factor in the long-term as well as the short-term demands).

However, it is not the role of the market actors to decide what costs and benefits are to be included in a market price; instead, those cost-benefit decisions are shaped by government and typically by legislatures. Imperfections, such as externalities, are the rule and not the exception; indeed they are to be expected of a system where imperfect political markets inevitably lead to imperfect legislative solutions that then impose imperfect institutional frameworks to underpin the economic markets. Only a political authority can correct these market frameworks, and this in itself should warn us that externalities will never be eliminated. Thus a market economy should be presumed to contain distortions that range from small to large, and even "extra-large." Furthermore these distortions can range far beyond asymmetries of information to include asymmetries of power and their routine abuse.

Thus, it really matters to think of the economic markets of capitalism as part of a system of political economy and not just one of economics. My conception of capitalism broadens the focus from market operations to include both the institutions that shape the market frameworks and the political authority that designs as well as governs the institutions in which markets are embedded, and thus encompasses political economy rather than the narrower notion of pure economics. In proposing this conception, I aim to suggest that the evolution of a capitalist system is as much a political phenomenon as an economic one, and specifically that it requires the visible hands of political actors exercising power through political institutions, such as elections and legislatures, in activities that are remarkably different from the unguided or invisible hand that Smith so astutely recognized.

The following monograph will proceed as follows: After a brief overview and critique of current conceptions of capitalism, I delve into the details of my own. I organize this discussion around the major characteristics of capitalism: (1) Capitalism is an indirect system of governance; (2) capitalism is analogous to organized sports; (3) capitalism is comprised of three levels—markets, institutions, and political authority; (4) the third level of political authority underscores the role of visible human agency, not just that of invisible market forces, in capitalism; (5) the political authority has the administrative opportunity and arguably the responsibility to shape the capitalist system to favor certain interest groups over others, as well as the entrepreneurial responsibility to modernize the capitalist system over time; (6) capitalism is a system of governance not only for private goods but also for public or "common" goods, where some of the most important of those common goods are the market frameworks themselves and where political authority, not market forces, is essential for governing the latter; (7) political authority inevitably shapes capitalism according to a strategy, no matter how implicit or imperfect that strategy might be; and (8) political and economic markets determine the nature of political authority, such that the political system of governance and the economic system (i.e., capitalism itself) are not only interdependent but also a theater of competition in which economic and political actors compete with each other for power.

I conclude with a summary of the work, restating my definition of capitalism in more simple terms and suggesting directions for the future study of capitalism as well as implications for contemporary society.

Chapter 2

Historical Conceptions of Capitalism

Historians, most notably Fernand Braudel with his three-volume *Civilization and Capitalism*[1], have traced the origins of the term capitalism to the mid-1800s. However, its notoriety came a few decades later, from socialists who used it as a term to describe what they disliked about the workings of liberal markets. Karl Marx, arguably one of the most prominent socialists of the time, used it as a way to refer to a system of markets that in his view favored capitalists at the expense of society.[2] His notion was, of course, conditioned by historical experience up to his own time as well as his own perspective on that history; when he was writing, markets appeared to inevitably pit capitalists versus the proletariat, without much regard for the fact that a democratically elected government, or even a limited monarchy, might intervene to protect the interests of the middle classes let alone the poor. In his era in both the United States and Europe, capital was achieving extraordinary power for newly emergent industrialists. For example, the largest firms in the US grew from perhaps 100 employees in 1800 to more than 100,000 a century later, and they grew still more in terms of the financial and physical resources at their

---

1 Fernand Braudel, *Civilization and Capitalism, 15th-18th Century, Vol. II* (Berkeley: University of California Press, 1992), 237.

2 Michael Merrill, "Putting 'Capitalism' in its Place: A Review of Recent Literature." *The William and Mary Quarterly*, 3rd Ser., Vol. 52, No. 2. (April, 1995), 315-326: 322. Footnote 21 gives a very lucid description of some of the history, though mostly in a US context.

command.[3] This extraordinary accumulation of private power called for a new conception of capitalism; Adam Smith's conception of atomistic capitalism, where firms had little or no economic power, was hardly an adequate framework for such analyses. At the same time, there were virtually no large-scale democratic states until almost the end of the 19th century; Britain enlarged its electorate from about 1.5% of its population to 2.5% in 1832, and then only by the late 19th century began to add wealthy merchants and manufacturers to its class of wealthy aristocrats. The US was the outstanding exception, as Alexis de Tocqueville recognized during his first hand study of the US in 1830.[4] But the fact that governments had not mounted much by way of any of successful attempts to embed markets in regulatory frameworks to protect labor, a critique brought up by Karl Polanyi, did not mean that they could not do so, as Marx implied, but only that it had not yet done so.

Despite its grounding in a particular historical context, Marx's critique became an influential understanding of capitalism during the mid- 19th century, and his ideas served as a sort of handbook for revolutionary activities. Notably, they provided a covering ideology for those who wanted to establish totalitarian regimes to suppress the power of the capitalists in a perverted recipe that allowed a few to govern in the name of the proletariat, while in reality they were not held accountable to anyone. In such a context, capitalism was hardly a term of approbation. Indeed this competition for ideas and for power was clouded by the fact that capitalism had been defined by its adversaries more than by its proponents; proponents

---

3 Alfred Chandler, *The Visible Hand* (Cambridge: Belknap Press of Harvard University Press, 1977), 50-65, 204. Alfred Chandler reports that prior to 1840, few firms employed over 50 workers, but that by 1891, the Pennsylvania Railroad Company alone employed over 110,000 and, indeed, it was not even the largest US railroad in terms of mileage at the time.

4 Alexis de Tocqueville, *Democracy in America* (New York: Alfred A. Knopf, Inc., 1945).

were preoccupied with resolving differences between utopian views, such as those of Robert Owen, with the near opposite view of laissez faire capitalism, a view that assumed market outcomes were based upon a system that predated government and were therefore not to be disturbed by government, with rare exceptions. The democratic alternative to both sets of views had yet to show much either in theory or practice that it could form market frameworks meeting Polanyi's challenge.

This very brief introduction to the history of capitalism in the 19th century is only intended to suggest that, by the late 19th century, it was a rather imperfect alternative to feudalism, in fact creating a new order that was open to huge concentrations of power that simply replaced those of the earlier order. Thus, ironically, capitalism came to be defined by some of its critics as the rationale for creating a centrally planned, coercive state that would monopolize power even more than its feudal predecessor. Although the democratic capitalism that we tend to take for granted today already existed in a few places, such as the United States, its existence was pushed into the shadows by the obvious presence of the new industrial giants, even in the United States, in the mid 19th century. Democratic capitalism has been challenged almost since its inception by oligarchic capitalism, though in the US case this challenge was delayed for some 200 years or more.

Over the last century and a half, the prevailing conception of capitalism has undergone a rather remarkable evolution, in terms of both its inherent structure and its impact on societal outcomes, both of which are of very direct import for this discussion. A century or more ago, the notion that markets were political as well as economic constructs was obvious; indeed, economics was then called political economy. At the same time, capitalism was a little used term, except as an epithet by its critics. Since then, economics has gradually narrowed its focus from political economy to economic relationships. From there, the focus has narrowed further to economic relationships that can be mathematically modeled, as though economics were a science devoted to the discovery and exposition of a system of natural laws. From this narrowed perspective, microeconomics has become the study of how markets – traditionally, the essential institution of capitalism – coordinate decentralized decision-making through a price mechanism to bring supply and demand into equilibrium without any explicit human agency or planning. Economic actors are presumed to interact on the basis of rational self-interest in a largely self-regulating economic system controlled by the laws of supply and demand. Rational self-interest is presumed to be universal and context-free (not to mention bubble-free), as are the laws of supply and demand. And capitalism, though based upon property rights created by human agency, is presumed to be able to achieve optimal outcomes for society without the benefit of explicit human agency, as though markets were controlled by natural forces akin to those of a gravitational field, a claim that might have been plausible in Smith's era, but surely is not in our own. In the terms to be used in this monograph, microeconomics and the prevailing conception of capitalism are now largely focused on markets alone. As this market-based conception of capitalism is one with which my definition most strongly contrasts, I find it appropriate to describe and then critique it here, before introducing my own view.

Chapter 3

Some Current Conceptions of Capitalism: Discussion and Critique

As I suggest above, many economists and even many historians today tend to equate capitalism with markets, and the activities of market actors alone. Capitalism, for them, is a system of natural forces, i.e., supply and demand, that naturally tend toward equilibrium. Notions of governance, let alone government, have little if any place within this impersonal, "scientific" system and are, in fact, often accused of corrupting or distorting capitalism. The strength of this market-based conception of capitalism has been apparent for many years, as highlighted by social historian Michael Merrill over a decade ago. In a 1995 review of contemporary conceptions of capitalism, Merrill pointed out the prevalence of the market-based conception and the challenges inherent in overcoming it: "If capitalism is little more than a synonym for a market economy, then any opposition to capitalism necessarily becomes an opposition to markets – in other words, an opposition so rarified and unreasonable to most people as scarcely to matter historically…"[5] But such opposition is crucial, he asserts, because capitalism is not simply a product of economics but of *political economics*.[6]

Thus far, I agree with Merrill. However, I feel that his argument ultimately disappoints in that he does not propose a sufficient alternative con-

---

[5] Merrill, op. cit., page 317.

[6] Merrill, op. cit., page 317.

ception. For him, "Capitalism, properly speaking, is not just an economic system based on market exchange, private property, wage labor, and sophisticated financial instruments…Capitalism, more precisely, is a market economy ruled by, or in the interests of, capitalists."[7] This second conception, while somewhat of an improvement on the first, fails to capture the actuality of capitalism in two key ways. First, it assumes that the interests of capitalists not only *do* prevail but *should* prevail in any capitalist system; it overlooks the possibility and even the desirability of governing markets in the interest of society as a whole. Second, it presupposes a notion of governance without explicitly recognizing the actual roles that human agents from the political sphere must play in a capitalist system if the market frameworks are to reflect the public interest through proper recognition of true social costs and benefits. Thus, Merrill leaves us with a critique of the market-based conception of capitalism without effectively moving beyond it. By tagging on the notion of the power of so-called capitalists, Merrill seems to be placing the theory of market-based capitalism in the context of what he sees in his own contemporary society; he does not ask if the contemporary context may be aberrant. But Merrill is certainly not alone in providing a clear and pointed critique, yet a less-than-robust alternative. Others have been equally unsuccessful in challenging the prevailing conception of capitalism, and, I believe, it is not least because this conception has been so effectively put forth by economists over the past half-century, and notably by Milton Friedman, whose work I review below.

The work of Milton Friedman, a Nobel Prize winning economist who became famous as a leader of the so-called Chicago School, is perhaps one of the most important representatives of today's market-based theories of capitalism. In his much-cited book *Capitalism and Freedom*, Friedman takes a more focused and less historical perspective of capitalism than I

[7] Merrill, op. cit., page 317.

do; he emphasizes the coordination of economic actors through voluntary bilateral transactions in a marketplace. Friedman states that the main theme of his book is to elaborate on "the role of competitive capitalism – the organization of the bulk of economic activity through private enterprise operating in a free market – as a system of economic freedom and a necessary condition for political freedom."[8]

In his conception of capitalism, i.e., competitive capitalism, Friedman is primarily focused on trade, and he is much concerned about political freedom. He points to the economic freedom of markets as essential to its political equivalent, a proposition that finds strong support in the political science literature. However, for Friedman, political freedom seems to mean the absence of coercion of one individual by others:

> "The fundamental threat to freedom is power to coerce, be it in the hands of a monarch, a dictator, an oligarchy, or a momentary majority. The preservation of freedom requires the elimination of such concentration of power to the fullest possible extent and the dispersal and distribution of whatever power cannot be eliminated – a system of checks and balances. By removing the organization of economic activity from the control of political authority, the market eliminates this source of coercive power. It enables economic strength to be a check to political power rather than a re-enforcement."[9]

Essentially Friedman defines freedom as *freedom from coercion by others*, and implicitly assumes that those "others" are political actors and not economic actors. In other words, Friedman assumes that only government can concentrate enough power to threaten the freedom of individuals; the concentration of power in the economic realm, such as by giant firms, and its threat to the freedom of individuals, such as that of smaller firms or the

---

[8] Milton Friedman, *Capitalism and Freedom, 40th Anniversary Edition* (Chicago: University of Chicago Press, 1962), 4.

[9] Friedman, op. cit., page 15.

employees of firms of any size, are omitted from his analysis, except for monopolies. Friedman overlooks the power of one firm to coerce another; when he assumes that competition eliminates economic power, he seems to overlook the fact that firms with thousands of employees compete with others that may have only one hundred employees or perhaps only ten. To speak of the transactions between giant firms and small ones as voluntary and without coercion seems quaint, almost as though it could be used to describe bilateral encounters between a whale and a school of minnows, from the whale's point of view. Relative size does not necessarily equate with relative power, but to ignore the potential for unequal power relationships in the private sector, and to focus only on its exercise by political authorities, seems a considerable oversimplification.

Overall, Friedman simplifies the reality of economic "freedom" by omitting consideration of the meaning of freedom to those members of society with relatively less economic power than others, in terms of meager resources, little education or human capital, and/or no financial capital with which to take advantage of market opportunities. Friedman seems to assume that inequalities in economic power are adequately controlled through competition, so long as most of the firms are privately owned, and that it is therefore only explicit inequalities of political power that must be avoided at all costs.

In point of fact, power relationships among individuals are rarely equal and, among various firms or teams of economic actors, even less so. Those with greater economic power can employ it as they bargain in markets or lobby political actors, while using even more overt coercion in less organized settings. In this more realistic perspective, economic power can be a force for the subversion of equality among persons, and thus a force for the subversion of freedom and democracy. To be compatible with democracy, and thus with the freedom of which Friedman conceives, capitalism needs to be modified or transformed in some way, as Robert Dahl has written.[10]

Modern economics has begun to recognize the narrowness of Friedman's vision, incorporating a notion of transformation into the study of capitalism. Specifically, in recent decades, formal economics has extended its field of study beyond markets to include the identification and examination of the institutional foundations of capitalism. Douglass North, a professor of economic history and recipient of the 1993 Nobel Prize in Economics, has been a pioneer in pointing out the need for such a change. As he implied in his acceptance speech: "There is no mystery why the field of development has failed to develop during the five decades since the end of the Second World War. Neo-classical theory is simply an inappropriate tool to analyze and prescribe policies that will induce development. It is concerned with the operation of markets, not with how markets develop."[11] North proposes a broader perspective, one that includes the forces framing those markets, i.e., institutions. He explains in his work that "Institutions provide the incentive structure; as that structure evolves it shapes the direction of economic change toward growth, stagnation or decline."[12] In recognizing that institutions *shape* the direction of economic change, Professor North implicitly recognizes that institutions shape markets in ways that can shape the behavior of market actors and eventually the path of economic growth as well. However, when he posits that institutions *evolve*, he does not take the next step to tell us *how* they evolve and whether their

---

[10] Robert Dahl, *Democracy and Its Critics*, (New Haven: Yale University Press, 1990). Dahl's writing here is further cited in Gabriel Almond, "Capitalism and Democracy," *PS: Political Science and Politics,* Vol. 24, No. 3, (Sep., 1991), 467-474: 470.

[11] Douglass North, Nobel Prize Speech. December 9, 1993. Published in *Nobel Lectures, Economics 1991-1995,* ed. Torsten Persson (Singapore: World Scientific Publishing Co. , 1997)

[12] Douglass North, "Institutions," *The Journal of Economic Perspectives*, Vol.5, No 1, (1991): 97.

evolution is a spontaneous process, like biological evolution, or one that is guided by human agency, like the construction of a road or a constitution.

What theories based on the work of Friedman and North miss is the idea of human agency in capitalism. True, the evolution of the institutions of capitalism is partially a spontaneous process that can spread gradually on its own, like increased sales and geographic distribution for a product or increased diversification in the output of a firm. But it is also partially an intentional process; unlike changes to sales that happen gradually and largely at the initiative of the firm, changes to the institutions that shape markets depend in large measure on political as opposed to economic choices, as when a state promulgates a new set of regulations that require changes in behavior from the economic actors. Friedman, North, and many others miss this notion of agency because they focus more on the trading paradigm of capitalism (i.e., private parties transacting business in markets) than on its production paradigm (i.e., private parties mobilizing resources to develop technologies in search of profits and thereby potentially exercising great influence over the direction of the markets).[13]

A brief elaboration of these two paradigms is in order here, such that the oversight of these economists is well understood. The trading paradigm can be broken down by the actors and forces involved, as follows: Private parties are allowed to transact business in markets, including entry into and exit from specific activities, while the price mechanism balances supply and demand, a framework of laws and regulations governs the competition, and an accountable government provides security, administers laws, and modernizes laws as appropriate. The production paradigm can be similarly characterized in terms of its primary actors and forces: Private parties are allowed to mobilize resources through various legal vehicles such as corporations to develop and exploit new technologies in

[13] For a similar perspective, arrived at independently, see Erik S. Reinert, *How Rich Countries Got Rich and Why Poor Countries Stay Poor* (London: Constable, 2007).

search of profits, while corporations are permitted to lock in shareholder capital indefinitely at the discretion of the board of directors,[14] and they are permitted the rights of self-governance through hierarchies; shareholders are shielded from losses through legislative grants of limited liability; managers are permitted to coordinate activities across functions and sectors through hierarchical organizations; employers are permitted to use implicit coercion, such as the loss of a job for employees who fail to carry out assigned roles; and competition for profits governs the allocation of resources and of internal rewards.

To ignore this second paradigm and see capitalism as nothing more than a system for trading, is to see it only with one eye, effectively overlooking what is arguably the greater source of the gains in technology and growth for which capitalism is known and, at the same time, the arena that is most susceptible to gross abuses of power.[15] Moreover, to overlook the production paradigm is to overlook the primary opportunities for human agency within capitalism. To explain: The trading paradigm requires institutions to play a supporting role in governing the markets in which trade occurs; the production paradigm, in contrast, requires them to play a much more active role in establishing and monitoring a decentralized system of private power and thus, in turn, further requires human agents to play a decision-making role with respect to the legal rights and responsibilities attached to such power. Put more simply, the former focuses on the product markets (i.e., for tradable commodities) by providing a framework in which to trade, while the latter focuses on the factor markets (i.e., for land, labor, and capital) by determining the relative mobility of resources and therefore the resulting distribution of power within the markets. To overlook the latter is to overlook the crucial processes through which capital-

---

14 Margaret Blair, "Locking in Capital: What Corporate law Achieved for Business Organizers in the 19th Century," *UCLA Law Review,* Vol. 51, No. 2 (2003): 387-455.

15 For a good discussion of this, see Reinert, op. cit.

ism actively evolves over time. For instance, absent the historical decision to reallocate the legal right to land ownership from established parties (e.g., the lords and the clergy) to a wider range of individuals, political, social and economic power would remain in the hands of a few, feudalism would persist indefinitely, and modern capitalism might never have emerged. Focusing on trade misses the importance of the production paradigm in developing the factor markets and thus in developing capitalism itself. Moreover, it misses the role of human agents, specifically political actors, in the emergence and ongoing evolution of capitalism.

As advanced as North's work is, relative to neoclassical economics, he joins Friedman in focusing more on trade than on production, more on the product markets than on the factor markets, and more on markets and their supporting institutions than on politics – and thus the human agency – shaping them. According to North, "The central issue of economic history and of economic development is to account for the evolution of political and economic institutions that create an economic environment that induces increasing productivity."[16] While true, accounting for the evolution of the institutions that enhance productivity takes one further into the realm of political science than North goes, examining how the capacities of governments are in turn influenced by political institutions that are quite simply outside of the purview of organized economics.

In a later work North does express greater awareness of the agency of the political realm in shaping the economic realm, specifically in terms of shaping property rights: "The efficiency of the political market is the key to this issue. If political transaction costs are low and the political actors have accurate models to guide them, then efficient property rights will result. But the high transaction costs of political markets and subjective perceptions of the actors more often have resulted in property rights that do not induce economic growth, and the consequent organizations may have

---

16 North, op. cit., page 98.

no incentive to create more productive economic rules."[17] Yet this analysis still reveals a narrow view of capitalism. North identifies a role for political agency, but circumscribes this agency as one driven by cost-benefit analysis; in other words, political markets exist, but they function as simplistically as Friedman's economic markets and lead to similarly simplistic outcomes of economic growth or decline. Inequalities in economic and political power, as well as their tendency to shape institutions and their outcomes, are completely missing from the picture.

North's work thus not only oversimplifies the evolution of institutions that enhance productivity, but also underemphasizes the idea that institutions can induce or reduce inequalities within society. The latter certainly merits attention, if only because excessive inequalities open the way for the empowerment of elites who can use their economic power to subvert legitimate government, especially if it is a nominally democratic government. As Tocqueville observed more than 150 years ago, most revolutions have been started either by people who wanted to reduce existing inequalities or, at the other extreme, to avoid their reduction, a pattern that further illustrates how a society's political and economic systems are inevitably intertwined.[18] Economic governance thus inevitably involves political institutions as well as political objectives, and capitalism cannot be reduced to the impersonal science of market forces alone.

To view capitalism as a system of governance, we must follow North's progress beyond Friedman to recognize the decisive role of institutions in shaping the markets, equilibrium or no. But we must then go further to recognize that the evolution of these institutions is in turn built upon human agency, as the political system determines the rights, responsibilities, and resulting powers of individuals and institutions within the economic

---

[17] Douglass C. North, Asbjorn Sonne Norgaard, and Richard Swedberg, *Institutions, Institutional Change and Economic Performance,* (Cambridge: Cambridge University Press, 1990), 52.

[18] Tocqueville, op.cit., page 611.

system over time. We must recognize that some of the essential coordinating processes that influence economic development lie beyond the traditional bounds of economic analysis, beyond the narrow, market-focused scope of neoclassical economics, and beyond even the broader scope of North and his colleagues in institutional economics. The study of capitalism is a study not of economics but rather of *political economy*, an interdisciplinary approach that prevailed until the emergence of neoclassical economics at the turn of the twentieth century and to which we must return today, if we are to truly understand and thereby shape our capitalist system of governance. Such is the motive behind my own conception of capitalism, discussed below.

Chapter 4

My Conception of Capitalism

My conception of capitalism is an attempt to address the above oversights in Friedman, North, and their colleagues' theories of capitalism. In conceiving capitalism to be a system of governance, I mean to move beyond neo-classical theory, where markets spontaneously coordinate the activities of economic actors through the price mechanism, to the broader form of analysis of political economy. Where North adds a second level of analysis involving the institutions that shape those markets with incentives and constraints, I am adding a third level where a political authority governs how those incentives are designed or shaped through a political process, and eventually administered.

My primary claim is that the visible hand of human agents in government is necessarily involved in establishing and maintaining the institutional structures that in turn shape the markets in which the invisible hand of the pricing mechanism operates. Capitalism can neither emerge nor develop without such constant human intervention. While it may be useful to speak of the emergence of capitalism by way of an "evolution" of human institutions, this evolution cannot be accounted for through the study of natural forces, as in biological evolution. Unlike biological systems which evolve through natural selection among random varieties, capitalist systems have been driven by human purposes from their very origins. Furthermore, they have the capability of purposive adaptation; they can take a step backwards in order to advance two steps forward at a later time, an act that cannot normally be achieved by a biological system. Such purposive adaptation implies a strategy, even if an imperfect or incoherent one, on the part of government, which in turn implies the existence of varieties of economic governance and thus varieties of capitalism.[19]

To put it simply yet clearly: Friedman conceives of capitalism as a one-level system for achieving economic coordination (i.e., economic markets), North conceives of it as a two-level system (i.e., economic markets embedded in institutions as shaped by cost benefit analysis), and I conceive of it as a three-level system (i.e., economic markets, institutions, and a political authority accountable to political markets).

Capitalism is an Indirect System of Governance

My historical and theoretical studies of capitalism led me to my own definition of capitalism as an indirect, three-level system of governance. I begin here by explaining its indirect nature. Capitalism is an indirect system of governance because the economic actors are governed by laws and rules that set conditions for acceptable behavior; it contrasts with two historical and two contemporary systems of governance for economic relationships. The first two are slavery and feudalism, both of which have become largely or completely obsolete, at least at the societal level. Slavery has a long history, was important as recently as the mid-nineteenth century, and figures strongly in the story of the early development of the western hemisphere. Feudalism, though largely extinct, has had a much more important role in economic history; indeed, capitalism emerged from centuries of feudalism in Europe in the period 1400-1800.

I focus less on these first two systems here because they are rarely found today and bear relevance primarily to the origins of capitalism, a

19 See Peter Hall and David Soskice, *Varieties of Capitalism* (Oxford: Oxford &University Press, 2000).

topic I only briefly address in this work.[20] Both slavery and feudalism preclude an essential ingredient of capitalism, i.e., relatively free factor markets. The defining institutions of capitalism tend to be in its factor markets (e.g., land, labor, and capital) and not in its product markets (e.g., fruits and vegetables, textiles and other traditional tradable commodities); the former are more deeply embedded in the political and social systems of a society and ultimately define how resources may and may not be used, not to mention by whom. More simply, products can be traded back and forth between anyone, even slaves, but the sale of land, the contracting of labor, and the lending of capital require a social system in which these factors are not fixed, as in feudalism or slavery. Some experts have referred to systems as capitalist because of the existence of small amounts of "free trade," despite the simultaneous existence of factor markets characterized by slavery, forced labor, or a feudal system where capital was not officially permitted to earn a return. I disagree; societies where forced labor or slavery are general conditions applying to a majority of the population do not meet the test of free markets, for both the products and the factors of an economy, that is essential to capitalism, no matter how much trade or entrepreneurial activity is engaged in by an elite few.

Though often left unsaid, and indeed unexamined, the freedoms of capitalism imply opportunities for personal growth and development. A system of forced labor or one with little or no educational opportunity for much of the population denies the substance of those freedoms to that fraction of its population. Historically speaking, those freedoms have been achieved primarily by overthrowing the prevailing social system; almost all "advanced" societies circa 1500 were governed through feudal systems, and the achievement of factor markets in these societies required a decisive break from feudal control of land and labor (e.g., the revolution in England

---

20 Please refer to chapter 5 of *Capitalism, Its Origins and Evolution as a System of Governance*, for a more thorough discussion of the origins of capitalism.

in 1689, that in France in 1789 and, later, the hostile takeover of parts of Germany and Italy by French troops). The crucial step in achieving capitalism in almost all countries throughout history has been the overthrow of the institutions of slavery and/or feudalism, liberating the factor markets for land and labor.[21]

The second two contrasting systems of economic governance still exist today. The first is a largely if not completely informal economic system where self-sufficiency, perhaps among family units, is practiced with only a modest degree of specialization or trade. In such cases, the rules for property ownership and trade are informal, and they depend upon a family or tribe as a coercive authority to enforce them. Historically, this latter form of organization characterized many indigenous peoples, and it still has scattered exemplars today. The second contemporary alternative is based upon direct control of human and other resources through a hierarchy backed by the coercive powers of a state, as in the former Soviet Union. It is against this alternative, one that has arguably been more prevalent since 1900 or so than the aforementioned three, that I most frequently contrast my conception of capitalism as an indirect system of governance. I refer to this statist alternative as a *direct* system of governance through a hierarchy, where governance can be by command and control. Likewise, I refer to my own conception of capitalism as an *indirect* system of governance, where governance occurs not by political authority itself but rather through the rules and institutions it shapes.

Figure 1 lays out the three contemporary economic systems, two in which economic coordination takes place under the auspices of the state

---

21 The achievement of capitalism in Australia, New Zealand and the US was arguably exceptional because feudalism never was strongly entrenched in these countries, and this was only slightly less true in Canada. Canada started out in a feudal land holding pattern along the St. Lawrence River, but most of its territory was developed under British laws following British takeover early in the 18th century. For much the same reason these same countries were also early to achieve democracy.

and a third where such coordination is entirely private and informal. The figure also identifies different forms of intervention and coordination. Adam Smith's invisible hand is one of the formal coordinating mechanisms in a capitalist economy, but only one of the three. I will explain the other two mechanisms shortly.

Milton Friedman, in his own work, and particularly in *Capitalism and Freedom*, correctly recognizes the economic systems in column one, while he mixes columns two and three of Figure 1. In my opinion he incorrectly claims that the system in column two is governed much like column three and is capitalism. Friedman identifies capitalism, i.e., competitive capitalism, as "The kind of economic organization that provides economic freedom directly… [and] also promotes political freedom because it separates economic power from political power and in this way enables the one to offset the other."[22] Friedman thus explicitly removes government as much as possible from his competitive capitalism, claiming that in this economic system, "an impersonal market separates economic activities from political views."[23] Government's role is to "determine, arbitrate, and enforce the rules of the game" of capitalism and not to directly participate in it.[24] And even in this supporting role, the government's role is to merely codify custom, or that which has already been agreed on: "most of the general conditions [of capitalism] are the unintended outcome of custom, accepted unthinkingly."[25] What Friedman is describing is, therefore, not capitalism but rather the informal system of column three:

---

22 Friedman, op. cit., page 9.

23 Friedman, op. cit., page 21.

24 Friedman, op. cit., page 27.

25 Friedman, op. cit., page 25.

Organizing authority:	*The state*		*Private parties*
Forms of state intervention:	Direct	Indirect	No formal role for the state
Planning:	Central plan with mandatory targets	Framework of laws and regulations establish a context for decentralized decision making	Decentralized decision making based on informal rules and understandings
Mechanisms for coordination:	Central plan plus state ownership and direction of enterprise	1. Pricing mechanism 2. Market frameworks 3. Corporate strategies	Informal pricing mechanism for informal commerce
Enforcement:	Enforcement of decisions through state bureaucracies, e.g., line ministries and central bank	Regulatory terms and conditions enforced by the coercive powers of the state (e.g., regulators and courts)	No legitimate coercive enforcement mechanism
Economic System:	Controlled economy	Capitalism	Unregulated trade
Examples:	The former Soviet Union	The United States	An informal market

Fig. 1. Three systems for organizing and coordinating economic activity

Friedman further diverges from Figure 1 by claiming that "there are only two ways of coordinating the economic activities of millions,"[26] and not three. He identifies the first as "central direction involving the use of coer-

[26] Friedman, op. cit., page 13.

cion – the technique of the army and of the modern totalitarian state."[27] This form of coordination corresponds to column one of Figure 1. Its alternative, according to Friedman, is "voluntary co-operation of individuals – the technique of the market place"[28], i.e., his notion of capitalism or column three of Figure 1.

Friedman's notion of capitalism is not capitalism but rather an informal market and, thus, a scenario that rarely exists today, at least in the more developed economies. Consider his elaboration on the coordinating force key to his capitalism, i.e., voluntary cooperation:

"The possibility of co-ordination through voluntary co-operation rests on the elementary – yet frequently denied – proposition that both parties to an economic transaction benefit from it, *provided the transaction is bilaterally voluntary and informed.* Exchange can therefore bring about co-ordination without coercion. A working model of a society organized through voluntary exchange is a *free private enterprise exchange economy*—what we have been calling competitive capitalism."[29]

In this key paragraph Friedman seems to forget what he says above, that government's role is to "determine, arbitrate, and enforce the rules of the game," all of which implies the right of a government to use coercive power and thereby place conditions upon the freedom of economic actors. Voluntary transactions, or trade, are indeed a crucial component of capitalism, but, like competitive play in an organized sport (an analogy to which I will return below), the economic actors in a capitalist system enjoy freedoms that are conditional, i.e., they are free to act only so long as they remain within the parameters of the laws and regulations that define the various markets of capitalism. If neither party can count on the state to use its coercive power to protect the respective parties from failures by the other

---

[27] Friedman, op. cit., page 13.

[28] Friedman, op. cit., page 13.

[29] Friedman. op. cit., page 13, Italics original.

party to deliver the goods, services, or payments due, on time and in the conditions agreed to, then this commerce is not capitalism. It is nothing more than informal trade and belongs in column three.

The purely private means of coordination that Friedman refers to, via a price mechanism and interpersonal trust between the participants, is not, strictly speaking, even part of a capitalist system. Instead, these relationships define the informal economy; it is a grey or black market where there are no formal standards to define what is being traded or who has what rights and responsibilities before or after the transaction. It can exist alongside a capitalist system, but, as anyone from a developing country or one recovering from civil war knows, such a lack of formality reduces the efficiency and transparency of market transactions. Imagine how much business would be transacted by credit cards if there were no security in their use and no recourse in the event that a card were lost or stolen. US law requires the issuer to be responsible for all such costs in excess of $50, and the issuer can recover such charges by spreading a small insurance cost across all users. That compulsory responsibility for the issuer to be responsible for such charges is part of the genius of capitalism; the issuer is much better placed than the user to stop such losses rapidly and to eventually collect any damages.

Informal commerce may be said to be "free," but this freedom is not the orderly commerce that is the hallmark of capitalism. Instead, it defines the uncertainty and disorder of a free-for-all. The farm stand selling fresh vegetables to a random passerby may seem like a good model of informal, voluntary capitalism, but if it does business without a zoning permit and periodic visits from the health authorities, it is most likely an illegal operation that may also be a source of unsafe produce. Moreover, absent rules enforced by a vigilant political authority, it may conduct business in a way unsafe to not only private but also public goods, as in polluting the environment. This issue of public goods or, more simply, the "common," will be addressed again later in this monograph.

In contrast to Friedman, I conceive of capitalism as the indirect system illustrated by column 2 of Figure 1 and simplified below in the right-hand panel of Figure 2. All formal markets are governed by laws and regulations, and these laws and regulations must be backed by the coercive powers of a legitimate political authority (typically its state bureaucracy) if they are to constitute effective frameworks for market transactions. Coordination within this formalized system is achieved by three mechanisms, the first of which is the price mechanism. The second is the whole institutional framework that underpins these markets; it is the administrative apparatus through which the visible hand of government translates estimated societal costs and benefits into various rights, taxes, and subsidies in order to approximate true social costs for each particular society. The third mechanism is private, but it is not based upon voluntary actions by consenting adults like informal markets are; it depends instead upon the strategies of firms, especially large firms, and upon hierarchical control exercised within those firms.[30]

Capitalist governance thus stretches beyond the bounds of economic markets to include the institutional foundations that both underpin and shape those markets. It is neither an informal nor a direct system of governance, but rather an *indirect* system of governance. As the informal or tribal system does not exist on any large scale today, contrary to the implications of Friedman's assertions, the most relevant alternative to capitalism is that of direct governance, and it is primarily that with which I contrast it from now on in this work. The contrast between direct and indirect governance can be highlighted by contrasting top-down governance through hierarchy within a firm and indirect governance through rules and regulations as practiced in organized team sports, as shown in Figure 2.

---

30 Please refer to the appendices of *Capitalism, Its Origins and Evolution as a System of Governance* for a more detailed discussion of the role of firms in a capitalist system.

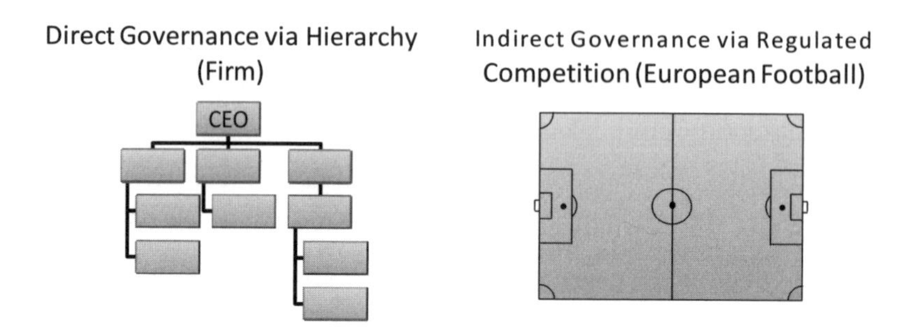

Fig. 2. Direct and Indirect Governance

Capitalism is Analogous to Organized Sports

The sporting analogy introduced above proves useful as I go further in explaining my conception of capitalism and contrasting it with that of others, such as Friedman, and I will accordingly refer to it often. As suggested in Figure 2, capitalism parallels organized sports in that the institutional context shapes but does not directly control the behavior of the actors. To continue with the sporting analogy, if the institutions of a football contest mandate a ball that is round and prohibit the use of the hands except in a few tightly defined circumstances, then the players can be expected to compete like European footballers. Put those same players in a game with an oblong ball and permission to use their hands to throw and catch it, and they are apt to play like American, Australian or Canadian footballers. The institutional context of organized sports shapes the behavior of athletes, but it does not directly control their behavior, and this parallels the governance of capitalism.

Before delving into details, it is fitting to note that my use of organized sports as an analogy to capitalism is not unique. Specifically, Friedman also employs this analogy, comparing the "day-to-day activities of people" to the "actions of the participants in a game when they are playing it" and likening the "general customary and legal framework" within which these activities take place to "the rules of the game they play."[31] However, as with his conception of capitalism in general, his analogy contains some significant oversights; he omits any specification of a political process for its governance or any notice of cumulative advantages in capitalism. Instead, he emphasizes the voluntary nature of submission to rules and conditions in both sports and society. While asserting the need for agreement to the rules or conditions, as well as the need for a system of arbitration (e.g., an umpire by means of a government), Friedman ascribes the source of these rules and conditions to custom or general consensus and claims that "no set of rules can prevail unless most participants most of the time conform to them without external sanctions."[32] How does this fit with market frameworks that may have been established through a 51-49 vote in a legislature or perhaps sustained by a 5-4 vote of a court? This reality seems far from custom, consensus, and un-coerced conformity.

Friedman's use of the sporting analogy further falls short in overlooking the political complexities of the process of establishing and reforming the rules of the game, capitalist or otherwise. According to Friedman, "These are the basic roles of government in a free society: to provide a means whereby we can modify the rules, to mediate differences among us on the meaning of the rules, and to enforce compliance with the rules on the part of those few who would otherwise not play the game."[33] I agree with this list as far as it goes, but, in explaining it, Friedman fails to tell us much of anything about what shape the market frameworks take, which in-

[31] Friedman, op. cit., page 25.

[32] Friedman, op. cit., page 25.

[33] Friedman, op. cit., page 25.

terests they might favor, where the rules come from, or how they are modernized.

Friedman here falls short both in his theory's statement and in its practical application. First, he puts forth a contradiction. As mentioned above, Friedman claims that the market of competitive capitalism "separates economic activities from political views."[34] Yet at the same time, Friedman acknowledges that "the role of government…is to do something that the market cannot do for itself, namely, to determine, arbitrate, and enforce the rules of the game."[35] If the rules of capitalism are created and modified by political actors, how can they ever be devoid of political biases? Moreover, for Friedman to assert a clear separation between economic and political power is to contradict not only his own theory but also reality; the market frameworks, i.e., the laws, are always created by political actors and therefore, to some extent, always contain a political agenda or tilt within them. To say, as Friedman does, that the rules of capitalism are the unintended outcome of custom, formalized through government, makes it sound as though the outcomes are almost as obvious as natural laws where *most* participants *most* of the time conform. Friedman in fact almost asserts as much, stating that in the impersonal markets of competitive capitalism, "no exchange will take place unless both parties do benefit from it."[36] How does he then account for situations where some parties benefit far more from an exchange than others, as when a CEO receives a large severance package to step down from a position of failed leadership, or for situations where some parties use their economic power to force economic exchange, as when employees must accept reduced wages in order to retain their jobs? While one or both may be welfare enhancing when viewed narrowly, as separate transactions, their broader effect is likely to legitim-

---

34 Friedman, op. cit., page 21.

35 Friedman, op. cit., page 27.

36 Friedman, op. cit., page 13.

ize opportunistic behavior that helps lower the behavioral expectations of all.

Friedman's conception of a competitive capitalism where markets are impersonal, apolitical, and unbiased, and where government plays as minimal a role as possible, is not the capitalism that we live today or, arguably, any sort of capitalism that has ever existed. It is a conception of an informal economy, something that exists only in the world of the fruit-and-vegetable stand or the black market, where exchange is predominantly governed by trust between local participants. Friedman's notion is, in a way, a return to Adam Smith's 18th century notion that government has discharged its responsibilities simply by supplying "easy taxes and a tolerable administration of justice,"[37] dismissing government as too unimportant to merit attention. Smith's position was much more adequate in his time frame than it is today, as his was a world largely though not exclusively characterized by atomistic competition. Today's large firms have perhaps 1000 times the employees of their counterparts in his days, and still more relative to his notional pin factory. Today, Smith's position attracts the loyalty of many who like its ideological implications (i.e., little to no government "intervention" in the markets) and are able to overlook the hugely transformed circumstances of the large firms that emerged late in the 19th century and whose descendants are still with us. And, unfortunately, it is a position that helps sustain those who wish to use economic reasoning to help refashion the law, as though atomistic competition were still the norm, and thereby becomes a form of political reasoning masquerading as economics.

To understand capitalism as organized sports in the way that Milton Friedman does is to overlook the essential roles of institutions and government. Moreover, to use these ideas as a basis for deregulation of a modern economy is to open wide the gates to a free-for-all in the markets. As

---

[37] Adam Smith, as favorably cited by Gregory Mankiw, "Repeat after me," *The Wall Street Journal*, January 3, 2006.

studies of capitalism throughout history may illustrate, economic actors, if left "free" to exercise their powers in a so-called free enterprise context, can challenge and even overwhelm government, thereby suborning democracy in favor of oligarchy. The study of US capitalism is particularly revealing. The industrial giants of the 19th century US economy grew in terms of employment roughly 1000 fold during that century, and surely far more than that in terms of the assets and income streams that they controlled. Furthermore, these same giant firms took over much of the coordinating function from the markets, as Alfred Chandler pointed out in his justly famous masterpiece *The Visible Hand*.[38] US authorities, judicial as well as political, permitted a vast growth of power in private hands, while at the same time reducing the accountability of US firms to those same political and regulatory authorities. Reducing the role of government in this case and many others throughout history can lead to an extreme and arguably unjust concentration, and then abuse, of economic power.

Government is not alone in its capacity to allow the abuse of power by others or by itself; given a chance, the private sector can and often will abuse its powers so that markets work for the few and not the many. Governments must restrain and regulate those with private power if they are to fulfill their responsibilities to protect the citizenry, i.e., to provide tolerable law enforcement. In addition, and as discussed further in this work, capitalism requires that government play a positive role in providing the public goods for which it is responsible and without which most people cannot expect to take advantage of the opportunities that capitalism can provide. From this enumeration of the essential roles of government, it should be clear that the political realm cannot be cleanly separated from the economic and that any analogy of capitalism with organized sports must therefore take into account the role of governing authorities (e.g., league organizers, referees and judges).

---

[38] Chandler, op. cit.

As a final critique, Friedman's comparison of capitalism with organized sports overlooks the key area where the two in fact differ. Namely, power earned in the economy can be used to influence political decisions and, in this respect, capitalism is quite distinct from and perhaps weaker, as a system, than organized sports. In organized sports, the teams are normally of equal size, much as in the model of atomistic competition. However, in capitalism, one firm may be ten times, a hundred times, or even a thousand times the size of another. Capitalism can thus support oligarchy, even a corrupt oligarchy, and in such a case it is not the guarantor of the freedoms that Friedman, with his simplistic sporting analogy, claims it to be. Those freedoms cannot be expected to be secured unless civil society is alert to the unequal distribution of power within the system, particularly that of political power. Laws do not make themselves nor enforce themselves. Unless there is demand for enforcement, it will not normally happen. On all of this Friedman is silent. The issue of power relationships in a capitalist economy is key to my conception of capitalism, as well as to the implications that may be drawn from it, and thus I will take the latter portion of this monograph to better establish the scope and interplay of the linkages between economic and political power.

Having outlined and critiqued Friedman's analogy of capitalism and organized sports, I now return to finish the outline of my own, aiming to improve upon the weaknesses identified in his as well as elaborating upon my conception of capitalism as a three-level system.

Capitalism is a Three-Level System of Governance

All organized sports can be understood as three-level systems, as suggested in Figure 3. The first level is the game itself, in which athletes compete with one another, whether as individuals or as teams. This competition is usually the focus of audience attention, watching to see who wins or loses as well as how the game is played. But while this competition appears to be primary and all that is really needed for the game to take place, it is in fact far from sufficient. Organized sports are not played in back alleys or out in the tall weeds, nor at random times among random assortments of athletes. Rather, the actual competition usually unfolds in carefully marked-out areas, at specific times, and under the supervision of a set of referees. The use of an explicit setting, set of rules, and team of regulators for sports parallels capitalism's nascent beginnings in the late middle ages, when it was confined to specifically designated market locations and market days and was often carried out according to a prescribed set of rules, often under the direct supervision of duly chartered guilds of registered tradesmen. Organized sports thus parallel the organized economic activity of capitalism. Likewise, unorganized sports (e.g., throwing around a football according to few to no rules that are determined and enforced by the players alone) parallel an informal economy, or column 3 of Figure 1 above.

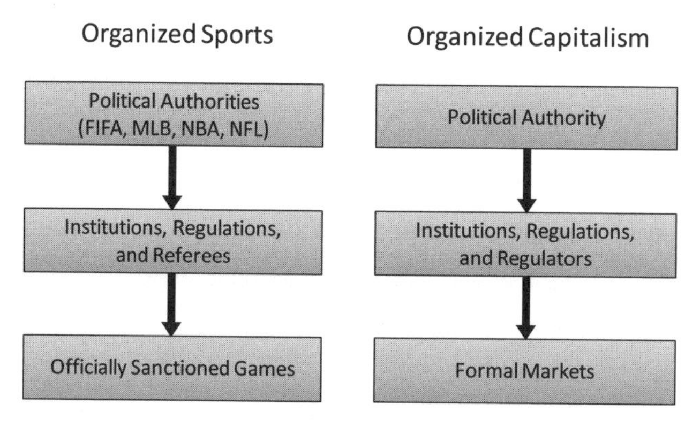

Fig. 3. Organized Sports vs. Organized Capitalism

In organized sports, the players, coaches and other team personnel comprise the first level of the system; the boundary conditions for such a contest are created and maintained by the administrative and regulatory officials who comprise the second level. More specifically, these agents demarcate and indeed maintain the field, specify the rules of play and the scoring system, and monitor the play. These agents organize and legitimate the competition and ensure that it is carried out on a level playing field, with no unfair advantages permitted.

These institutional foundations (e.g., the officials and the rules they enforce) are in turn created and legitimated by a governing organization comprising the third and final level. This governing organization is a political authority with the power to decide on the rules, i.e., who is eligible to compete, the time and location of the games, and technologies that may be used. In professional sports, the political authority may also have the power to set the terms and conditions for the distribution of certain revenues among participating teams, a power that can be exercised to limit disparities in incomes by team, thus curtailing the relative power of one or a few teams to dominate the sport year after year.[39] For instance, the Olympics are organized as individual sports under the auspices of an umbrella organization, a slight variant from the diagram above. International football is organized in the usual pattern, where the International Federation of Amateur Football, or FIFA as it is known in French, establishes the rules and hires the judges to monitor competition. US professional football is organized under the auspices of the National Football League (NFL) in a similar structure.

Thus, through this analogy to organized sports, I arrive at my conception of a three-level system of governance in capitalism. Capitalism is an indirect, three-level system of governance, where a political authority permits economic actors to mobilize and employ resources in competition with one another, subject to a set of laws and regulations as defined and enforced by one or more regulatory agencies. The political authority com-

---

[39] In the United States, the National Football League is widely recognized as the most socialistic of the organized sports because the league authorities have the power to distribute the television revenues approximately equally among teams despite the difference in the markets that they directly serve.

prises the top (third) level in the system; the regulatory and other institutional foundations provided by that political authority comprise the middle (second) level; and the regulated competition among economic actors in markets comprises the bottom (first) level.

On the first level, firms compete to secure their labor and capital as well as to serve their customers. In this competition, as with sports, individuals and firms mobilize and apply energy to achieve their goals, some following distinctive strategies while others will play it safe with a "me too" strategy. On the second level, the basic institutional foundations, including physical and social infrastructure as well as the individuals and organizations operating them, set the terms for the behavior of the actors on the first level. Physical infrastructure includes, among other things, transportation and communications, while social infrastructure includes the educational, public health, and legal systems. Those operating these basic institutional foundations and enforcing their rules are typically agents of the state, including specialized regulators who oversee behavior in certain industries. Examples include those who deal with food, drugs, or transportation and those who protect societal resources, such as the physical environment or safety in the workplace. On the third level, a political authority—typically one with specialized functions such as executive, legislative, and judicial branches—actively oversees and shapes the operations of the first two. A set of political institutions connects the political authority to the political markets (e.g., elections, which may be more or less democratic) and eventually to civil society, to which such an authority is ultimately accountable.[40]

---

40 Please refer to chapters 3 and 4 of *Capitalism, Its Origins and Evolution as a System of Governance* for an elaboration of my theory on how the economic and political systems connect.

Chapter 5

Political Authority Shapes Capitalism with Visible Human Agency

The distinguishing contribution of my theory of capitalism is the third level of political authority. Recognizing the role of human agents, those within government in particular, is central to articulating a more accurate theory of capitalism as well as to understanding the realities of capitalist societies.

As the theories of North but especially of Friedman demonstrate, capitalism is often defined without a notion of human agency let alone government. Most commonly, capitalism is understood in Friedman-like terms as the process by which economic markets utilize the "invisible hand" of the price mechanism to spontaneously coordinate supply and demand between actors competing for particular goods and services. However, in practice, the visible hands of human agents are implicated in the process as they guide the invisible hand of the pricing mechanism. Specifically, the invisible hand can only align individual and societal priorities if the institutional foundations of capitalism have *shaped* those markets so that individual costs and benefits reflect those of society rather than those of an unruly mob or powerful elites. The pricing mechanism cannot come close to achieving an optimal coordinating role absent the effective work of the visible hand of government, often through legislative processes such as a parliament.[41]

Followers of Friedman tend to not only overlook but also actively reject this role of government in a capitalist system. According to them, informed, voluntary and bilateral transactions are the essence of a self-regulating capitalist system and therefore that system can and must be free from governmental coercion. In reality, coercion is to be found in most capitalist markets; large firms coerce those that are smaller, a patent holder enjoys market power, an employer typically authorizes only one employee to make a job offer to a prospective employee, and employees may or may not organize to bargain in a similar format. The regulatory institutions deal routinely with various forms of coercion; it is the rule and not the exception. Likewise, these institutions themselves employ coercion to create the freedoms of a capitalist system. Quite unlike Friedman's theory of the almost absolute freedom of economic actors, the reality of a capitalist system demonstrates that the so-called freedom of economic actors is almost always *conditional*, and conditional not so much on the voluntary actions of a trading partner as on rules and regulations established by the state. Successful capitalism depends not only upon the state granting power to private actors to enter, compete in, and exit from markets, but also upon the state restraining private actors so that they do not abuse this power. In a capitalist system, the participation of private economic actors depends on their agreement to follow the rules set and enforced by the state. Capitalist freedom is thereby conditional, and political authority shapes the conditions to ensure fair play among competitors who have very different powers.

---

41 The emergence of capitalism does not require democracy; in fact, it appears to be a necessary precondition for the latter. Please see chapter 5 of *Capitalism, Its Origins and Evolution as a System of Governance* for more explanation on this point.

Thus, when we look at the reality of capitalist systems, we see that while ostensibly free competition in economic markets is an essential and utterly distinctive feature of capitalism, it represents only part of a capitalist system and not the totality. Yes, capitalism relies upon the concept of competition to energize human actors and on prices to coordinate their actions. But it also relies on a notion of regulation to limit or constrain the behavioral practices in which economic actors are permitted to engage. For instance, every transaction is subject to a process of authentication, a process necessary in the event of a dispute over what was agreed to.

Regulation and authentication of the terms of a trade cannot be accomplished by the economic actors alone; their isolated, individualistic positions in economic markets prevents them from effectively, not to mention legitimately, adjudicating differences or enforcing settlements. A higher-level entity with legitimate authority and coercive power is needed, i.e., a political authority. Barter can take place in a back alley, with no authentication of the transaction and no records. But organized capitalism, in contrast, requires the auspices of a political authority to create and legitimate one or more regulatory authorities to authenticate transactions, adjudicate differences, and coerce enforcement when necessary. As a result, markets are embedded in, or underpinned by, institutions that are in turn parts of systems of law, public administration and ultimately government. The design and shaping of those institutions, the monitoring of the behavior of the economic actors within them, and the application of coercive force to demand remedial behavior if and when needed are all crucial functions of the political authority within any system of capitalist governance.

Recognizing a strong role of government in the shaping of institutions that in turn shape markets and indeed economies has in fact proven crucial to economic development. For instance, the timely modernization of capitalist institutions in Europe was an important contributing factor to the ability of certain states to finance military forces to maintain independence and ultimately survive the nearly continuous warfare of the region in the critical formative centuries. Countries, such as England and the Netherlands, that had overthrown arbitrary rule by divine right, could raise money more easily and at much lower rates of interest than those ruled by absolute monarchies, as in France and Spain; the former were thus able to hire mercenaries to protect themselves from repeated attempts of hostile takeover by the latter. In such circumstances, preservation of autonomy came ahead of efficiency as a political and economic goal, and achievement of this goal required maintaining an active ability to mobilize resources rather than passively accepting resource mobilization as beyond societal control.

As the above historical example demonstrates, economic governance implies far more than facilitation of equilibria in markets; it implies that markets are achieving the purposes for which they were designed, which initially included helping to finance warfare. Mercantilist policies were not necessarily the folly that they are sometimes supposed to have been; they were a means to mobilize economic power to help maintain independence. Whether wise or not, states had policies other than promoting consumer welfare, and they used human agency in the design of economic institutions to pursue societal objectives, actively directing the markets towards their desired equilibria.

While market based coordination has proven a very desirable and efficient system much of the time, equilibrium itself is not necessarily a sufficient indicator of appropriate governance. Economic markets can achieve undesirable equilibria which yield extraordinarily distorted circumstances. For example, they can even achieve so-called equilibrium during a crisis, as the system implodes, as in the cases of the Great Depression and of the more recent chaos following the bankruptcy Lehman Brothers in mid-2008. Indeed, as another example, the US mortgage markets were able to achieve remarkable efficiencies as well equilibria from the mid-1980s onward, thanks to new financial engineering. However, these equilibria were accompanied by (or rather, supported by) distorted circumstances. These included the inflation of a bubble, as both consumers and lenders took on reckless levels of leverage, in the expectation of continuously rising prices, and as US financial leverage relative to GDP almost tripled between 1980 and 2006, making all such transactions far more risky. The distortions seemed to have escaped the attention of US regulators as well as that of the economic actors themselves, leaving them unchecked until the emergence of the credit crisis in August 2007, a crisis that gradually became a full-fledged, worldwide economic crisis in 2008. All throughout this process, the invisible hand of market forces struggled to maintain equilibria in supply and demand across thousands or likely millions of economic actors in more than 100 countries and currencies. But the invisible hand could not judge the adequacy of the design of the market frameworks in which transactions between those actors took place and, as is now clear, many and perhaps most of the economic actors and regulators proved equally inept at judging the adequacy of the system. Many were even arguably unwilling to engage in such judgment, believing instead that any outcome of a so-called free market system would be acceptable if not ideal. To ignore the fundamental import of the regulatory role of a political authority in such circumstances is to substitute ideology for analysis and to invite chaos, as the results of their inaction now demonstrates.

As the above examples indicate, the basic premise of this work is that capitalism is not simply a system of economic relationships that are coordinated through the invisible hand of the pricing mechanism in markets; it is also, and perhaps more importantly, a system of governance that requires first, the articulation of political vision to guide the design market frameworks that will work toward achievement of societal goals and, second, the mobilization of political power to implement those frameworks so as to shape the markets, to monitor the actions of human agents who ensure that the competitors follow the rules, and, crucially, to modify these institutional frameworks as needed to ensure that the markets yield results that are considered to be broadly in the interests of society. No invisible hand can create the frameworks in the first place, nor monitor them, nor to design and implement modifications to correct their unwanted side effects. The essential institutions of capitalism cannot develop along with the needs of society absent the informed and capable input of human agents, such as those empowered through a government.

Gabriel Almond, a professor of political science at Stanford University, stated these basic ideas of capitalist governance: "The economy and the polity are the main problem solving mechanisms of human society. They each have their distinctive 'goods' or ends. They necessarily interact with each other and transform each other in the process."[42] As problem-solving mechanisms, they require human agency within them. This essential human role means that capitalism is a mix of sociology, administration, politics, economics, and law and that any theory of capitalism must include not only an economic level and an administrative level, but also a political level, what I call here the third level of political authority.

Political Authority Plays Both Administrative and Entrepreneurial Roles in Shaping Capitalism

In sports, as indeed in capitalism, the level of the political authority encompasses two distinct roles: one administrative, in maintaining the existing system with its approved teams, rules, and existing organization for the monitoring and enforcement of the rules, and the second entrepreneurial, in mobilizing power to win the needed votes in the legislature in order to admit new teams, change the locations or timing of competition, change the rules and regulations, and/or change the distribution of revenues. Every time a political authority wishes to enact change, its leaders must mobilize enough power to overcome the forces that wish to protect the status quo. In organized sports, the political leaders may have gained their position of power by purchasing a league franchise to own a professional team. While

42 Almond, op. cit., page 467.

they typically operate through political bodies (e.g., an executive and a legislature), the members of the league's legislature own their seats and typically are not accountable to an independent electorate. In addition, the entrepreneurial aspect of teams exercising political power in organized sports is very different from that of firms exercising political power in democratic capitalism, insofar as the political authorities, for most organized sports, operate under a grant of immunity from antitrust laws, which allows them to govern their league through face to face consultation much like the officials who govern a state. Teams in a sports league can sit together as a legislature to revise the rules of play, admit a new team to the league, and even legislate a split of revenues, if they wish (e.g., television revenues). In contrast, firms can mobilize lobbying power through trade associations but are not usually permitted to control entry to their industry or to split revenues, let alone rig prices; the governing political authority alone may legitimately take on this more entrepreneurial role.

With this comparison in mind, consider the dual role of political authority in a capitalist system. The continued success of capitalist systems depends upon the periodic modernization of the legal and regulatory frameworks as indicated by changing market conditions and societal priorities. Government therefore must play the two distinct roles, as administrator and as entrepreneur. In the short term, quasi-static perspective, government and its agents administer the existing institutions, both physical and social. In a longer term perspective, government must have the capacity to modernize these institutions as conditions indicate. This second role requires both foresight to recognize needs and entrepreneurial skills to mobilize enough power to effect the needed changes through a legislature. Given its added complexity of intentional change (as opposed to simple maintenance of the status quo, in the administrative function), this second role requires further elaboration.

Successful capitalism requires a system of governance that is built from the premise that there is no one solution or fixed set of relationships that is best for all times and circumstances. The system of governance must be able to manage its own capabilities and legitimacy in such a way that it can be efficient in meeting its responsibilities in the short run, while evolving as circumstances change, including as societal priorities change, so that it can cope effectively over the long run with a changing context. This means that government must be entrepreneurial, identifying changes that need to be made and mobilizing the political power to effect such change in legitimate ways and in a timely fashion.

Adam Smith had one of the great insights of all time when he recognized that the invisible hand of the pricing mechanism could coordinate economic transactions in ways that spontaneously served the public interest. But when he opined that "Little else is required to carry a state to the highest degree of opulence from the lowest barbarism but peace, easy taxes and a tolerable administration of justice,"[43] he overlooked the entrepreneurial role of government in the continuous process of modernizing the institutions of capitalism. Can we honestly say that the "tolerable administration" of laws and institutions is remotely adequate to meet the entrepreneurial responsibilities of government in the appropriate shaping and periodic modernization of laws and institutions in a complex society? For instance, is the formulation of patent laws to protect inventors as well as investors and consumers a matter of tolerable administration? Or is the development of food and drug regulations to protect patients and to provide due process for speeding new products to market any less a matter of genius than how to price automobiles, promote the sales of soup, or to educate consumers to new varieties of hair spray or deodorant?

Smith had the genius to recognize that markets could coordinate the actions of disparate actors in ways that might be superior to explicit bureaucratic planning (like that attempted in France under Jean Baptiste Colbert in the era of Louis XIV) as implied in his famous passage introducing the

---

43 Adam Smith, as favorably cited by Gregory Mankiw, "Repeat after me," *The Wall Street Journal*, January 3, 2006.

invisible hand: "As every individual...endeavours...to employ his capital in the support of domestic industry, and so to direct that industry that its produce may be of greatest value; every individual labours to render the annual revenue of society as great as he can. [While] he intends only his own gain...he is in this, as in many other cases, led by an invisible hand to promote an end which was no part of his intention."[44] Smith did not claim that markets are *always* right, but he did imply that they were right in this and "many other cases."

Are there examples where they can coordinate the actions of many buyers and sellers in ways that are not in society's best interests? Consider a real estate market in which housing prices are rising, down payments on mortgages are unregulated, and real interest rates are zero or even negative. Such a scenario occurred in the US in 2005-2007, under the watchful eyes of a Federal Reserve that seemed certain that the markets would sort things out, without added regulation. As is now evident, individual decisions in this market, characterized by rising assets prices and easy credit, were apt to involve cumulative speculation and the creation of a bubble. Smith's great insight implicitly assumes that the market frameworks take appropriate account of all societal costs and benefits. How can they take account of a speculative bubble that was facilitated, if not caused, by reflationary monetary policies managed by the Federal Reserve Bank, other than to wait for the bust? This would seem to be a situation that calls for human agency, in this case from the same Federal Reserve Bank that helped to facilitate the growth of the bubble in the first place. Moreover, it calls for agency on the part of the prevailing political authority (i.e., government) and its regulators taking an entrepreneurial, active role in economic affairs.

Choosing (or not choosing) to adapt to a changing context by in turn changing the institutions and market frameworks of a capitalist system im-

---

44 Adam Smith, *Wealth of Nations*, Oxford World's Classics edition (Oxford: Oxford University Press, 2008), 291-292.

plies not only political agency in promoting general social interests, as above, but also political agency in economic development. This returns us to the notion of the production paradigm of capitalism, introduced briefly above. Capitalist development is built from profit opportunities for investment in new technologies and markets, in a context where the opportunities induce increased supply and therefore competitive pressures that generate a Darwinian selection process which weeds out ineffective uses of societal resources. In order to facilitate such investment, capitalism – via the level of political authority – allocates legal rights and responsibilities to permit the existence of different forms of organizations that can exercise differing powers and accept differing risks. For example, governments can permit the chartering of joint stock companies as a legal vehicle for the mobilization of capital, companies that can have a life independent of their founders. Historically, this was a huge departure from the prevailing partnership form, which, in some countries, had to be reconstituted whenever a partner died or retired. Governments can also distribute risks in different ways among various economic actors through the institutional frameworks that it creates and legitimates, while at the same time allowing the economic actors themselves to decide how to share the risks and the rewards of economic transactions within those frameworks. For example, the institution of limited liability for shareholders shifts some of the risks of failure from shareholders to creditors, thereby making it easier for entrepreneurs to raise capital, a key consideration in promoting economic development. Not least, the modern corporation has the power to solicit private investment on terms where it need never return the money to the investor; the latter can recover part or all of his funds only by finding a third party to purchase his shares at the going price. In permitting such power to mobilize and lock up capital, governments may entrust great power to private parties in the hope that the firms will use this power in ways that contribute to the general economic development of their respective societies, as well as those societies' general socioeconomic well-being.

Finally, and on a more macro-level, recognizing an entrepreneurial role for government not only implies recognizing that there is agency on the part of government, but also implies recognizing that there can be a number of varieties of capitalism. The varieties reflect political choices that in turn reflect societal preferences. Capitalism is based upon a generic concept of indirect governance, but there can be different societal preferences that affect the outcomes towards which government indirectly shapes various markets. For example, a societal choice to prohibit collective bargaining, a choice many societies have taken in their early development, is implicitly a strategy to favor capital over labor, at least until such time as the society has become more prosperous. Likewise, a societal choice to prohibit one firm from buying shares in another is implicitly a strategy to try to build a society that has few large centers of private economic power.[45] Perhaps the most significant of these societal choices, enforced (and at times distorted) by government, is how to handle the public goods or common resources of society, and it is the topic to which I now turn below.

Capitalism is a System of Governance for Public, Common Goods as Well as Private Property

So far, our discussion of capitalism has generally focused on private goods and actors. But capitalism is a system of governance for public goods as well, from the environment, to a system of defense, to the law, to the institutions of capitalism and democracy themselves. Collectively, these tangi-

---

[45] The US was such a society until the 1880s, when New Jersey's legislature authorized such purchases by firms domiciled in New Jersey. Chapter 13 of *Capitalism, Its Origins and Evolution as a System of Governance,* discusses this example in further detail.

ble and intangible resources can be understood as the common property of society or as "the common" for short.

Any such common is not likely to survive without government recognition and support. Specifically, in the informal, voluntary situation of Figure 1, column 3, none of the parties would be in any way obliged to pay their fair share of the costs for the use of the common institutions and resources provided by society through government. They would be free to refuse to pay taxes while despoiling the common, much as US chemical, oil and steel companies did when using the rivers as sewers early in the twentieth century. By the same token, a druggist could sell ineffective or indeed dangerous drugs, such as opium or poisons, and an auto manufacturer could sell "lemons," without fear of customer recourse. This is precisely where capitalism (Figure 1, column 2) achieves huge gains in transparency and effectiveness compared to a less formal if superficially freer system. Simply put, a legitimate political authority employing coercive force indirectly through politico-economic institutions ensures that such abuse is limited or perhaps non-existent.

To understand the importance of regulating common goods and particularly the commercial common of capitalist institutions in general, consider the general contrast between an unregulated system and a regulated one (e.g., Figure 1 columns 3 and 2, respectively) in history. The traditional common was a pasture where a number of farmers or shepherds could share the right to graze their animals, and it had little by way of a formal structure of governance. Absent a political authority to ensure such governance, it was difficult to get the economic actors to limit their usage of the common and even more difficult to get them to accept their fair share of the responsibilities for its maintenance, let alone its improvement. Thus, an inadequately regulated agricultural common could be abused by some of the actors; for example, some might allow their animals to over-graze and damage the land to the disadvantage of all. Moreover, even without such intentional abuse, the lack of a system for improvement could limit any gains in productivity from this common resource.

As an indirect form of governance, capitalism creates a somewhat similar common, i.e., the commercial common, where many actors have rights to compete for access to a set of resources and also for the right to sell into a set of markets, all in a context where other actors have similar rights and responsibilities. However, this commercial common has a different history than that of the traditional common, having its origins in a governance system for a much less tangible resource, the market frameworks themselves. Formalization of rules has been crucial to the development of this common resource, which might initially entail little more than providing a legitimate source of authority to enforce a set of rules for the trade of goods and services, as already agreed upon by the economic actors themselves. Over time, this commercial common gradually and naturally took on a physical as well as intangible reality as it became important to have roads for travel, designated places for trade, physical protection of the economic actors from thieves, perhaps including unscrupulous tax collectors, and a legal process for adjudicating disputes.[46] However, this commercial common became something quite different when it was extended from the product markets to the factor markets, i.e., the markets for land, labor, technologies and capital. , and thus fully become part of a capitalist system.

As suggested earlier in this work, the deepening of the commercial common to include the factor markets typically required dramatic changes in power relationships, for example to free serfs from their feudal obligations and allow them instead to work for wages. The same was true for freeing land from feudal contractual obligations and for obtaining permis-

---

[46] Garrett Hardin, an eminent biologist, wrote a famous paper on "The Tragedy of the Commons," only to recognize later that the tragedy came not from the concept of the common per se but from the lack of effective regulation in how it was used, maintained, and developed through time. See Garrett Hardin, "The Tragedy of the Commons," *Science*, New Series, Vol. 162, No. 3859. (Dec. 13, 1968), 1243-1248.

sion to amass power through legal vehicles such as firms. Such deepening of the commercial common to include the factor markets generally required violent change, through conquest or revolution. As a result, it did not happen gradually the world over but rather in some locations centuries before others. The experiences of North America and South America in the period 1500-1800 are a particular contrast.[47]

As the commercial common has been established in different regions over history, its regulation has been a critical issue. Prior to the advent of long distance trade, circa 1500, people all over the world were able to manage their various local physical commons because those commons were small enough for the actors to see the damage that resulted from over-hunting or over-grazing. These actors would then govern themselves accordingly and maintain a stable system whose output was limited. Opening relatively isolated communities and markets to trade and specialization led to the destruction of many such commons and to a loss of social cohesion in those smaller, more rustic communities. A similar problem remains today, albeit on a much larger scale. Successful globalization depends upon successful regulation of a global common, including successful regulation of atmospheric pollutants and of the harvesting of marine life. While excessive regulation has stifled many economies for long periods, inadequate regulation is also a threat to effective decentralized decision-making throughout the global common. Abuse of the common is an ever-present temptation that comes with economic freedom. Effective use of a commercial common, as well as its effective protection from abuse, depends upon the maintenance of an effective system of economic governance, and, for all practical purposes, today that means governance through a capitalist system headed by a legitimate political authority.

---

47 Please refer to chapters 6 and 7 of *Capitalism, Its Origins and Evolution as a System of Governance* for a more in-depth account of these contrasting histories.

Of course, the political authority regulating the common could permit abuse, either passively or actively, by the way it chooses to regulate (or not regulate), much as is suggested in the previous discussion of the state's entrepreneurial role. First of all, capitalist systems typically rely on the state to make direct provision of certain public goods, including highways, schools, and law enforcement, while refraining from the temptation to own, operate, or directly control the economic actors. But if the state does become a direct economic actor, for example as the owner of large enterprises, it becomes a player as well as a provider of institutional foundations of the system. This puts state agents in a direct conflict of interest; in terms of organized sports, they become players on the field with regulatory powers and thus only questionably subject to the discipline of the rules they set. There are times when it is appropriate for states to play both roles, as in the case of a national emergency or natural monopoly, but these interventions are best pursued for reasons of state, e.g. national security, and for a limited duration. If direct interventions are widespread and/or last indefinitely, they invite corruption and the distortion of market frameworks for the benefit of the few at the expense of society as a whole. In a second, more passive form of abuse, the government may indirectly contribute to others' abuse by allowing those economic actors with greater economic or political power to influence its own agents and thereby shape the institutions and markets of capitalism to their private advantage. Such indirect abuse (i.e., abuse via private actors), can, in fact, occur legitimately via an inherently corrupt strategy, as I will discuss below.

Political Authority Shapes the Economic System in Accord with a Strategy, Implicit or Explicit

As explained throughout this work, government's primary mode of intervention in a capitalist system is indirect, through the formulation and enforcement of the laws and regulations that guide behavior and through the provision of certain common resources. Nevertheless, the actions of government inevitably imply strategic "tilts" to the various market frameworks; they can tilt toward capital or labor, investors or creditors, producers or consumers, and so on. The market frameworks are shaped or tilted by government, and that shaping can be based upon quite different policy priorities, from protecting the status quo to promoting growth and development. These same market frameworks can accept more or less risk as well as more or less tilting for or against particular classes of economic actors (producers or consumers, etc.). Governments specify the responsibilities of the various participants in these transactions (e.g., for the safety and serviceability of the products) as well as the conditions under which they are produced and distributed. In short, political decisions by government inevitably induce the mobilization and/or allocation of societal and economic resources to favor certain interests over others.

For this reason, the actions of government, whether indirect or direct, inevitably imply a strategy, though this strategy is often largely implicit rather than overt. The strategy (or strategies) may broadly affect the economy as a whole or take a more tailored approach to affect a sector or subsector. The strategy may not be explicit, optimal, or even coherent, but it will inevitably favor some interests over others. In addition, however broad or narrow its scope and whomever it favors, the strategy is typically created gradually over time rather than as an immediate grand plan, and typically involves the inputs of many people with competing ideas. It may even be impacted indirectly or covertly by actors outside of the official public realm, such as by private interests sending campaign donations to favor one politician or piece of legislation over another in the hopes of promoting their private gain. Such interaction between the political and economic systems will be addressed in the following section.

A particular strategy takes effect at the most general level with respect to market frameworks. As the previous section on the commercial common suggested, market frameworks are key to capitalism; their shape and integrity determines the system's shape and integrity. Market frameworks define what property is and what rights belong to its owners. They define permissible behavior as the economic actors interact with one another, such as prohibiting price fixing but allowing discount pricing. And they define which elements of the physical and social infrastructure may be used in common by these actors as foundations for their activities, whether in production or trade. For example, if one is considering bidding for an empty piece of land, the bid price will be influenced by the market framework as well as by the bids of other actors, as the market framework spells out what property rights go with the piece of land. Can one build upon it? Can one only build a residential structure, or would a commercial or industrial structure be allowed? Can one build right to the official edge of the lot, or is there a minimum setback? Can one build to any height, with any mass in terms of cubic feet, and with any architectural style? Are subsurface mineral rights included? The applicable property rights are established in the market frameworks through the decisions of various political authorities. In the case of the US, they are most often established by local zoning boards, while in Europe, they are typically established and administered by provincial or even national governments. These rights affect the potential value of the property for all bidders, and the bidding takes place within the framework that is so established. Having that framework, local (as in the US) or national (as in Europe), can therefore make a great difference in terms of the relative power of the competing bidders. Furthermore, government, at whichever level, also reserves the right to change the frameworks from time to time as societal priorities change.

Government can exercise strategy in its tilting of market frameworks not only with respect to property rights but in the product as well as factor markets. Consider first two examples from the product markets, namely those for gasoline and pharmaceuticals. The significance of differing frameworks is well illustrated by the contrast in prices between these two markets in Europe and the United States, as shown in Figure 4.

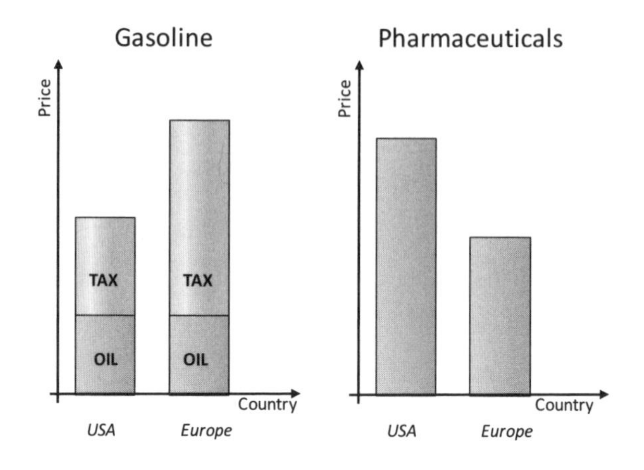

Fig. 4. Market frameworks, for instance in the product markets, differ from one country to another

The differences in gasoline prices are accounted for largely by differences in excise taxes among the various countries, with the United States levying a much lower tax on gasoline than its European counterparts. These differences derive from the countries' contrasting strategies: The Europeans have used the gasoline tax as a source of general revenues, while the US has from the beginning earmarked gasoline taxes primarily for highway construction and maintenance. Moreover, as a by-product of these differences, the Europeans have relied on gasoline prices to induce more efficient automobiles while the US has, with less success, attempted to reduce gasoline consumption by establishing regulatory standards of fuel economy for various classes of cars and trucks. Thus even when overarching goals may be similar (i.e., enhanced fuel economy), their achievement may differ significantly depending on which strategies are employed to promote them (i.e., a more market-oriented approach by Europe than by the United States).

When it comes to pharmaceuticals the story is roughly the reverse. The facts are stark: The US, virtually alone among developed countries, allows market pricing for drugs while most other developed countries have price controls. This difference in pricing policies by country has led many European pharmaceutical firms to shift important parts of their research activities to the United States, where they have a much better opportunity to recover their research investments. In a sense, then, US consumers end up footing much of the bill for pharmaceutical research for the rest of the world. At the same time, the US has developed a health care system where much of the cost is borne by employers. European competitors have an advantage in that their firms do not have comparable health care costs because the latter are mostly borne by their respective governments.

All these distinctions derive from the countries' contrasting strategies and resulting shaping of market frameworks. The United States, with strategies aimed both at promoting research and at privatized health care, has ended up shaping its frameworks to create a market where health care costs are much higher and borne disproportionately by employers. Many European countries, in contrast, have embraced a strategy of limiting health care costs and nationalizing their burden; the effects have thus been quite different than in the US, with pharmaceutical companies moving research abroad and consumers receiving widespread, typically universal and low-cost health care. In this case we again see how different strategies lead to very different market outcomes and thus very different varieties of capitalism.

Perhaps the most striking example of government strategy, in terms of its fundamental importance to modern capitalist systems, is that of government intervening in the factor markets, such as actively encouraging the mobilization of capital. Consider the various legal institutions government must set up to allow a corporation versus an individual to successfully (i.e., safely) invest his or her capital in a project and even borrow additional money to achieve a larger scope or scale. If the project fails, the individual investor is liable for repayment of the loan, while a corporation that has been granted the right of limited liability is in a different and preferred position. The investors in such a corporation can lose their investment if the firm fails, but they are not liable for the bank debts at all, in contrast to the individual above. The creditors can only claim their respective shares of whatever remains of the assets of the firm. By choosing to allocate legal rights in this way, the government has implemented a strategy that distributes risks and rewards among various interests – corporations, individual investors, and creditors. Government thus becomes necessary not only for enabling the effective existence of important institutions such as "limited liability," "foreclosures," and even "loans," but also for tilting their structure to give certain interests a better deal than others, depending on the circumstances.

What this means, simply put, is that government can strategically shape a capitalist system towards one set of parties over others. It is, in fact, through the articulation and implementation of strategies that the government fulfills its entrepreneurial role (especially in the production paradigm of capitalism), updating institutions to fit changing political, economic, and/or social circumstances. In an ideal world, government would update or re-shape the system to promote the general interests of society, which are often determined through political markets (e.g., legislatures). More often, however, it falls at least somewhat short of this goal to favor certain private interests over others. At times, this failure may be the result of corruption, as private interests exercise influence over political actors such that the latter shape the system to favor the former.

It is at this point, as we examine this notion of strategy, that the analogy between capitalism and organized sports falls short. The two are similar systems in that they operate on three levels, but there are some crucial differences. Most of these stem from fundamental differences in the purpose of the respective systems and thus become quite evident in consideration of the notion of strategy here.

The purpose of organized sports is to facilitate periodic competition among athletes, whether as individuals or as teams, both to encourage and recognize athletic excellence and to provide entertainment for the public. To this end, each sporting contest starts anew, teams are of equal size, and the advantages gained by a team during a game or a season are forfeited at the end of that game or season. In addition, and crucially, the entry of new teams is controlled by a system of franchises that may only be granted by a sporting authority; this authority in turn acts under an antitrust exemption and thus has sovereignty over its sporting league, like a state. Moreover, this authority, acting through referees and other agents, may intervene in the moment to regulate players and teams almost immediately, issuing penalties right when rules are broken or even putting players out of a game or season or disqualifying teams from participation; such immediate regulatory actions are designed to maintain a more level playing field.

In contrast, capitalism has quite a different purpose and set of regulatory capabilities. Capitalism, in its various forms, is intended to facilitate the productive use of societal resources in order to meet consumer needs in the short run and to raise the standard of living through time. As a result, its regulatory frameworks give priority to promoting productivity rather than the fine points of equalizing competitive resources on a given day or during a given season. At the same time, with rare exceptions, capitalism is regulated after the fact and not in real-time the way organized sports are. The regulators do not stop the play to assess a foul, nor do they halt the competition to examine a controversial event via "instant replay." The economy moves on, and disputes are settled after the fact, in court if need be.

The major contrast between organized sports and capitalism is that of a level, or not-so-level, playing field. While the institutions of organized sports are designed to ensure a level playing field, those of capitalism are not. To explain: Since economies of scale will enhance productivity, it follows that capitalism generally permits the accumulation of advantages, subject to certain exceptions and certain limits on acceptable behavior. It also follows that capitalism permits "teams" – i.e., firms – of radically different sizes to enter and exit industries without the approval of other participants, and it permits the entry of new competitors with new technologies that may give them an advantage over all other competitors. As a result, capitalism permits and encourages multifaceted competition among firms of different sizes using different resources on more than a single playing field (or industry) at a time. Capitalist competition, though regulated, is not designed to unfold between teams that are equal, nor circumstances that must be "level." Advantages, such as a playing field tilted in one's favor, become possible sources of additional – and potentially cumulative – advantages.

Since capitalism is designed to promote productivity, it can be expected to promote inequalities of income and wealth, and first movers in a technology may keep their advantages for decades. Capitalist competition is for keeps, not for sport. And it is up to the political authority to strategically shape or tilt economic institutions such that the unequal outcomes of capitalist competition do not ultimately undermine greater political, societal, or cultural priorities, priorities generally set or expressed by the political markets. If the playing field slants too much in the wrong direction, those out of favor may react negatively through the political markets (i.e., elections, referendums, or even revolutions), replace the current political authority, and thereby attempt to tilt the market frameworks in a direction more favorable to their interests and, hopefully, those of general society.

Chapter 6

—

The Political and Economic Systems Are Interdependent

As the foregoing section indicates and Figure 5 illustrates below, the political system and the economic system are inherently linked in a capitalist society; political markets influence the economic markets, and, in turn, the economic markets influence the political markets.

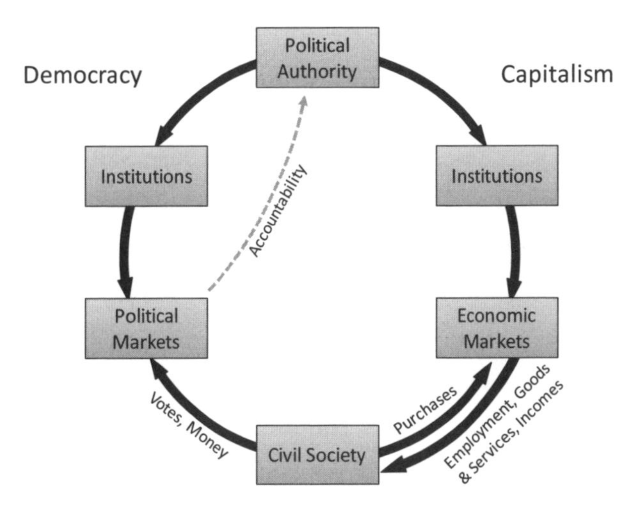

Fig. 5. Capitalism and Democracy are interdependent systems of governance

First, the outcomes of political markets directly determine the laws governing the economic markets and thereby indirectly affect their outcomes.

Specifically, legislatures are responsible for the design of the market frameworks in law, while regulatory authorities specify and interpret the regulations to implement those laws, with these sets of political and administrative actors legitimated by government. The strategic governance of the market frameworks can also be modified through the executive and judicial branches. While legislative, executive, and judicial forms of coordination are in this way all part of the governance structure of capitalism, it is the former that takes priority.

Legislatures are vital coordinating devices in a capitalist society; they bring different political actors or even political parties together to create compromises that, in theory, reflect conflicting interests and power relationships in order to achieve the public good. These compromises, again in theory, have taken into account all appropriate societal costs and benefits and thereby promote the interests of the middle class, in lieu of those of a wealthy elite or an aroused mob of the poor. In reality, this end is not always achieved, in that the political process does not always correctly perceive, reflect, or act in favor of true societal costs and benefits. Some costs are difficult to include, such as the costs of pollution, but typically because powerful interests resist their inclusion, through lobbying or other means; this is therefore not a market failure but instead a political failure. Other societal costs may be overlooked because politicians see them (or are encouraged to see them) as natural outcomes, such as the wages of low-skilled labor in a period of high employment and continuing immigration; this is, in fact, a political failure, given that the wage level is depressed not by the value of the labor performed but by weak bargaining power, a situation political actors could clearly mitigate. Supply and demand almost always reflect power relationships and thereby previous and current political considerations. Thus, in the reality of a capitalist system, the economic markets will not reflect society's true interests unless its political markets do.

But the political markets, as we have seen, can fail in this responsibility of identifying and balancing societal costs and benefits. The political markets follow no physical laws, like gravity or thermodynamics, but rather the actors' understanding of political as well as economic dynamics, an understanding that may be flawed (e.g., by poor research) or corrupted (e.g., by the financial influence of private interests). Could economic actors do a better job than political actors in evaluating the dynamics of the market and determining its frameworks? Perhaps, but likely not in favor of the public good and never with sustained legitimacy! Given their interested position within the market, any set of private actors would be biased in evaluating and then balancing societal costs and benefits. Moreover, these economic actors may well have the greatest knowledge of economic conditions and the most economic power, but they do not have the legitimacy to legislate or regulate beyond very narrow, mutually-agreed-upon limits. Only political actors do, though they may of course misuse their legitimate power, whether intentionally or not.

Economic markets are thus shaped by the (often imperfect) political markets of their respective legislatures. That shape is not set in stone; it can be changed by effective political pressures as well as intentional and unintentional asymmetries of information. Laws do not enforce themselves, and, in fact, they cannot even protect themselves from political pressures. Laws need continuous political support to survive. Thus, the system of economic governance is constantly and inseparably linked to the system of political governance, and it is the political system that has the legitimacy to shape the economic. However, the economic system is likely to have more information and a more targeted set of interests than the political system, so the agents of latter, i.e., politicians, are always likely to be in a position of trying to catch up to those of the former who have more knowledge and more money, i.e., businessmen. The political system will almost certainly make imperfect and sometimes even unwise choices, but it will in almost all cases determine the ultimate shape of markets, both economic and political.[48]

Ideally, the political markets are themselves shaped to minimize such imperfect and unwise decisions and thereby maintain legitimacy. Successful capitalism needs a government that is based upon the rule of law, but not necessarily a democracy. A government of laws depends upon the creation of checks and balances established through the structuring of the political authority of the state, e.g., its constituent branches (executive, legislative, and judicial) and levels of government (federal, state, and local), to ensure that the state does not encroach on the private spaces reserved for civil society. Early examples of effective governments based upon the rule of law include the limited monarchies of in Britain and of the Netherlands circa 1700, as well as the limited city based monarchies of Genoa, Florence and Venice, the latter pre-dating the year 1000. Ultimately, the alertness and civic consciousness of society are essential to ensure that its elected representatives limit the state's interventions in the marketplace and the temptations of state officials to claim an excessive share of privately earned gains.

I thus return to the theme of competition between the economic and political systems with which I began this paper. Any political system can expect competition between those who derive their power from the political markets and those who do so in the economic markets. Relative power of the actors will influence the tilt to market frameworks, directly by powerful political actors and indirectly by powerful economic actors influencing the former. Karl Marx supposed that liberal markets would be dominated by capitalists (i.e., powerful economic actors) and would lead to their do-

48 The United States is an important exception in that its Supreme Court can and has overturned legislative decisions that were explicitly designed to reshape market frameworks (e.g., those of the New Deal in the mid-1930s, elaborated on in my book, *Capitalism, Its Origins and Evolution as a System of Governance*).

mination of the political system as well. There was some truth to this at the time that he wrote, and it can certainly still happen today, but it is not a necessary outcome as he supposed. The tendency of capitalism to produce increasing inequality and eventually oligarchy over time is ever present, but it is a tendency that can be held in check, even if those checks are continually subject to challenge by would-be oligarchs.

When it comes to holding this tendency toward economic inequality in check and thereby maintaining a legitimate political system, I believe democracy has its advantages, given that it is based on government of the people by the people. However, even democratically elected legislatures are imperfect; for instance, a 51% majority may be enough to impose its will on the minority without much compromise. Moreover, legislatures, however democratic in nature, can still be vulnerable to the influence of moneyed private interests (as in the case of the U.S. both historically and today); a legislature that is dominated by concerns for the financial interests of the legislators can be expected to legislate for special interests and not for the people.

As Abraham Lincoln implied at Gettysburg, government *by* the people is no assurance that it is *for* the people. For the market frameworks of a capitalist society to best balance societal costs and benefits, the legislative markets must achieve outcomes that are both *by* the people and *for* the people. Thus we come to the crucial connection between the economic and political systems, or capitalism and democracy. Political leaders working through the political institutions of legislatures are responsible for shaping the institutions of capitalism such that the markets function *for* the people.

Chapter 7
—
Conclusions

Capitalism is a three-level system of indirect governance for economic relationships; it is a system that is political and administrative as well as economic. Organized markets cannot exist without a set of institutional foundations that establish various rights and responsibilities that are attributed to notions of property, and these foundations are created, legitimated, regulated, and periodically modernized under the auspices of a political authority such as a state. It is government and its agents, not private economic actors, who create and ultimately enforce the laws and regulations that guide production and trade. Since property rights are societal constructs and not gifts of nature, these rights will only take proper account of societal costs and benefits if they are established through a political process that is broadly representative of society itself, e.g., a democracy with a strong middle class.

Capitalism has three major coordinating mechanisms, and not just one. Two of the three depend upon human agency, while the invisible hand of the pricing mechanism works automatically. One of the visible hands belongs to government, and it guides the system, whether explicitly or not. The other visible hand belongs to the management of firms, and particularly large firms. Unlike government, the visible hand of management can coordinate product flows and financial transactions on a multinational basis.

As a "visible hand", government has two modes of intervention in an economy, direct and indirect. The indirect mode of intervention covers the maintenance and operation of the institutional frameworks that underpin all markets. It is essential to the operation of a capitalist system, not optional. The direct role is much more optional, for example in the ownership and control of public enterprises or the taking of land by the powers of eminent domain.

Government also has two quite different roles to play in any capitalist economy: as an administrator and as an innovator. The state bureaucracy takes on most of the responsibility for the administrative role, while political leaders must take on the responsibility not only for choosing the key administrative personnel making up the bureaucracy but also for recognizing the need for entrepreneurial innovation in institutions and for achieving them in a timely fashion.

If individual action is to add up to what is best for society, then the regulatory and other institutional functions of government, as it takes on its entrepreneurial role, must get the market frameworks right as well as securing the property rights of the economic actors. While there are no scientifically "right" answers in the realm of such governance, in democratic societies, at least, it is reasonable to define "right" loosely as the extent to which economic markets take appropriate countenance of societal costs and benefits. Achieving this depends in large measure on how well the political markets of its system of governance reflect societal interests. The central point here is that for economic markets to perform the coordinating function in the public interest, assuming this goal constitutes what is "right," the political markets of that society must see to it that legislature represents those interests and that its institutions work so that the outcomes are *for* the people and not just *by* their representatives.

Market forces alone cannot achieve these goals. One of the geniuses of capitalism is that markets tend to be self-correcting; excess supply leads to a decline in price and a reduction in supply. However, market frameworks are *not* self-correcting. Market frameworks have no way to correct their own imperfections, such as the under-pricing of pollution or the creation of excessive red tape. Only the intervention of the state can provide the necessary corrective measures to prevent capitalists or other organized groups from abusing the common for their own advantage and thus to promote the public interest.

Capitalism depends upon government to actively intervene in this way over time, managing and periodically modernizing market frameworks as circumstances change, including societal priorities as incomes rise. The appropriate modernization of market frameworks, including the tax and other policies necessary to avoid undue inequalities of wealth and power, requires the visible hand of government to make appropriate choices of policies and the mobilization of power for their enactment and administration. A society without effective government has little if any chance of progressing from barbarism to opulence for inequalities are likely to remain and worsen; such a society requires the visible hand of the state to intervene to modernize market frameworks in a timely way as well as to simultaneously administer and enforce existing rights and responsibilities as a complement to the invisible hand of the pricing mechanism in its coordination of the production, distribution, and trade of goods and services within its economy. In the long term, there cannot be effective capitalist development without effective governmental intervention to modernize its market frameworks in a timely and appropriate fashion. Effective capitalist development can be best assured by government that is for the people, which in turn requires political institutions that are fashioned to achieve such a goal. Government by the people may be essential but is surely not sufficient to achieve such a goal

The full responsibilities of government are not adequately recognized in the well-established, neoclassical view of capitalism. This view has tended to focus on the trading paradigm and the product markets of capitalism, and accordingly deemed their self-regulation by market forces sufficient and even ideal. It has tended to neglect the factor markets (i.e., the markets for land, labor, finance capital, and most recently for knowledge) of capitalism, markets whose advent and maintenance inherently require outside regulation from the political realm. Overall, this view has overlooked the necessary role of political authority and regulatory institutions in shaping economic markets, initially and over time, to promote certain societal interests over others. By the same token, this view has overlooked the notion of varieties of capitalism as expressions of the varying priorities of distinctive political entities and their associated societies. Thus, this view is a narrow and unrealistic conception of capitalism, one that inhibits not only an accurate understanding of how modern capitalist societies work but also the ability of leaders of those societies to effectively shape their market frameworks towards certain desired ends. It has also been associated with a Washington Consensus view of capitalism, as though there one best variety of capitalism that was associated with one particular nation state.

A broader definition of capitalism, such as that which I set forth in this work, makes room for a much broader understanding of the processes of economic development. It also makes room for varieties of capitalism, each tailored to its own circumstances, including its own societal priorities. The great strength of capitalism lies in its capacity to facilitate a society's development of and adaptation to new resources, technologies, and circumstances in general. But such powerful change cannot take place without leaders of a society, particularly leaders of its political authority, understanding capitalism and how actively they can and inevitably do play a role in its development..

Thus, I believe my three-level conception of capitalism offers an improvement over the prevailing neoclassical conception, carrying several implications for the study and practice of capitalism. To begin with, my conception implies a continuing role for human agency in economic development. It also implies the continuing evolution of institutions and market frameworks as well the evolution of supply and demand within markets. It recognizes that capitalism is far more than a science of how markets operate in that none of those markets exist absent the institutional foundations created, monitored, and periodically modernized by governments. It recognizes that without the essential and ongoing work of the visible hand of government to revise as well as enforce market frameworks, we would have much less developed capitalist systems, if capitalist systems at all.

Capitalism requires more than markets, firms, and individual economic actors; it requires structure, security, and adaptability that only government, in the form of human decisions, can provide consistently and accountably over time as circumstances constantly change. Until we accept government's framework-defining role as an essential feature, we will not have a satisfactory understanding of capitalism as a system of governance.

Epilogue — Why This Short Book and Why Now?

This book is the conceptual core of a larger work on *Capitalism, its Origin and Evolution as a System of Governance,* which will be published later this year. Due to recent events, a discussion about how capitalism actually works seems to be in progress in many countries, so I have decided to offer this shorter work as a contribution to that discussion.

The financial crisis that emerged in August 1997, as well as the situation we are facing today, was based upon a series of policy mistakes, and not on an "accident of history" such as a war or a stoppage of oil supplies from the Persian Gulf or a tsunami in the Pacific Ocean.[49] One of the principal theses of the larger book is that the series of policy mistakes are attributable in large measure on the one hand to various actors in the financial services sector, and notably to banks, "near banks", borrowers, and investors, and on the other hand to regulators and political authorities. These two groups of actors respectively engaged in and implicitly encouraged the new financial engineering that gave false confidence to borrowers and lenders alike and which contributed to acceptance of ultimately unsustainable levels of financial leverage. On the surface level, the common characteristics in these mistakes have included exuberant speculation, greed, and

[49] For more on this topic, please see David Moss. "An Ounce of Prevention: The Power of Public Risk Management in Stabilizing the Financial System." *Harvard Business School, Working Paper 09-087*. January 27, 2009.

inadequate understanding of complex securities. But at a deeper level, they have been built upon a fundamentally flawed understanding of capitalism, one that owes its origins to the writings and teachings of some well known economists. The central flaw in that understanding was the idea that markets can be, should be, and are self-regulating, as I point out in this work and its larger correlate. Markets alone do not make capitalism. There can be no large-scale, organized capitalism without effective regulation under the auspices of a legitimate political authority that has the coercive power to punish players who break the rules. The emergence of capitalism, as well as its evolution, always and everywhere has depended upon political decisions as a precondition for the legitimacy and, indeed, the existence of those economic markets. Additionally, the US economy in particular has been built on a distinctive set of premises that set it apart from most other countries, and it is not at all obvious that others should try to emulate the purportedly free market capitalism that was practiced from 1870 until 1937 and that has again been practiced in the US since 1980.

I believe a better understanding of capitalism would have helped regulators curb the excesses of the speculative behavior sooner. Moreover, going forward it will be very important in the design and legislative construction of appropriate remedies. This shorter work provides such an understanding, while the full length book builds further upon it. The full length book will explain the faulty model, its intellectual roots, and its intellectual sponsors. In addition, it will point a particular finger at a "toxic trio" of policies (i.e., reliance on self-regulation of the economic actors, the adoption of shareholder capitalism as a concept of corporate purpose, and the use of one-way, upside-only incentive compensation) that together set the stage for the disaster, a needless disaster that was allowed to incubate in an otherwise benign set of circumstances including high employment, steady growth and low inflation. It is my hope that these two works, the shorter and the longer, can contribute to the policy debate through the identification and explanation of a more adequate model of capitalism, one that cannot function on the basis of self-regulating markets any more than competitive team sports can function on the basis of unregulated competition. While capitalism is based upon the unleashing of human energies through markets, those markets, like their counterparts in organized sports, must be regulated by appointed officials who monitor the action and whose decisions have the backing of a political authority that has legitimate coercive powers to punish any actors who break the rules. Making the rules of capitalism is not up to the economic actors but to the political actors in their respective political markets, i.e., the elections and legislatures of their respective societies. So-called "free enterprise" is an unfortunate shorthand term in common usage today; the freedoms of capitalism are always conditional on obedience to a set of laws and regulation, and those laws and regulations emanate from government and not the private sector. To deregulate as a panacea, as occurred in the euphoria after 1980, and notably in Britain and the US, was a very unfortunate mistake. It led to speculative bubbles and then to chaos. Corrective action must be based upon reestablishment of effective regulation that is co-extensive with the markets. It is my hope that this short book and its longer correlate may contribute to the due debate on capitalism and the necessity for its active remodeling.

Printing: Krips bv, Meppel, The Netherlands
Binding: Stürtz, Würzburg, Germany